Lecture Notes in Physics

Founding Editors

Wolf Beiglböck

Jürgen Ehlers

Klaus Hepp

Hans-Arwed Weidenmüller

Volume 1029

The series Lecture Notes in Physics (LNP), founded in 1969, reports new developments in physics research and teaching - quickly and informally, but with a high quality and the explicit aim to summarize and communicate current knowledge in an accessible way. Books published in this series are conceived as bridging material between advanced graduate textbooks and the forefront of research and to serve three purposes:

- to be a compact and modern up-to-date source of reference on a well-defined topic;
- to serve as an accessible introduction to the field to postgraduate students and non-specialist researchers from related areas;
- to be a source of advanced teaching material for specialized seminars, courses and schools.

Both monographs and multi-author volumes will be considered for publication. Edited volumes should however consist of a very limited number of contributions only. Proceedings will not be considered for LNP.

Volumes published in LNP are disseminated both in print and in electronic formats, the electronic archive being available at springerlink.com. The series content is indexed, abstracted and referenced by many abstracting and information services, bibliographic networks, subscription agencies, library networks, and consortia.

Proposals should be sent to a member of the Editorial Board, or directly to the responsible editor at Springer:

Dr Lisa Scalone
lisa.scalone@springernature.com

Anders Malthe-Sørenssen

Percolation Theory Using Python

 Springer

Anders Malthe-Sørenssen
Physics
University of Oslo - Center for Computing
in Science Education
Oslo, Norway

ISSN 0075-8450   ISSN 1616-6361   (electronic)
Lecture Notes in Physics
ISBN 978-3-031-59899-9   ISBN 978-3-031-59900-2   (eBook)
https://doi.org/10.1007/978-3-031-59900-2

This Springer imprint is published by the registered company Springer Nature Switzerland AG
The registered company address is: Gewerbestrasse 11, 6330 Cham, Switzerland

If disposing of this product, please recycle the paper.

Preface

This textbook was developed for the course Fys4460 Disordered media and percolation theory. The course was developed in 2004 and taught every year since at the University of Oslo. The original idea of the course was to provide an introduction to basic aspects of scaling theory to a cross-disciplinary group of students. Both geoscience and physics students have successfully taken the course.

This book follows the underlying philosophy that learning a subject is a *hands-on* activity that requires student activities. The course that used the book was project driven. The students solved a set of extensive computational and theoretical exercises and were supported by lectures that provided the theoretical background and group sessions with a learning assistant. The exercises used in the course have been woven into the text, but are also given as a long project description in an appendix. This textbook provides much of the same information that was provided in the lectures.

I believe that in order to learn a subject such as scaling, the student needs to gain hands-on experience with real data. The student should learn to generate, analyze and understand data and models. The focus is not to generate perfect data. Instead, we aim to teach the student how to make sense of imperfect data. The data presented in the book and the data that students may generate using the supplied programs are therefore not from very long simulations, but instead from simulations that take a few minutes on an ordinary computer. The experience from this course has been that students learn most effectively by being guided through the process of building models and generating data. Some details of the computer programs have therefore been provided in the text, and we strive to use a similar notation in the computer code and in the mathematics in order to make the transfer from mathematics to computational modeling as simple as possible.

Another aspect of the book is that it tries to be complete in exposition and worked examples. Not only are the theoretical arguments carried out in detail. The computer codes needed to generate data are provided in such a form that they can be run and can generate the data in the examples. This provides students with a complete set of worked examples that contain theory, modeling (the transfer from theory to model), implementation, analysis and the resulting connection between theory and analysis.

In the full course, this textbook was only one half of the curriculum. For the first 10 years the first part of the course focused on random walks and the last part focused on random growth processes. For the second 10 years of the course, the course switched to be a course on cross-scale modeling of porous media. The first half of the course focused on molecular dynamics modeling of homogeneous systems in order to build an understanding of concepts from statistical physics from computational examples. The second part of the course used molecular dynamics simulations to model nanoscale porous media with focus on fluid transport (diffusion) and fluid flow in a nanoporous system and elastic properties of the porous matrix. Percolation theory was then introduced as a method to upscale the nanoscale systems, and we measured correlation functions, flow and diffusion problems across scales.

The course on percolation theory which formed the basis for this textbook was inspired by a course given by Amnon Aharony on random systems several times in the 1990s. This course was a great inspiration for all students and faculty and the course served as an inspiration for this course and for this text. Thank you for your great inspiration Amnon.

This book is written as a practical textbook introduction to the field of percolation theory with particular emphasis on containing all the computational methods needed to study percolation theory. Thus, we have included computer code where it is needed. The textbook does not aim to provide a complete set of references to percolation theory. Instead, only a few key references are included for students who want to explore more. There are many other good texts and reviews that provide a detailed set of references and a more historical description of the development of the field.

This textbook is the result of the contributions from many students in the course. Originally, the textbook was written with examples in matlab. However, as Python gradually have developed into the tool of choice for scientific computing, also the code in this course was updated. This was first done by Svenn-Arne Dragly, and some of the translations from matlab to Python was originally done by him. Later contributions from, e.g., Øyvind Schøyen on the translation of matlab to Python code for diffusion are also acknowledged. Thank you to all the students who have contributed in this course. It has been great fun to teach it because of your input and inspiration. I am greatly indebted to you!

Thank you also to my mentors Jens Feder, Torstein Jøssang and Bjørn Jamtveit who built up a cross-disciplinary research environment between physics, computer science and geoscience—the Center for the Physics of Geological Processes. You have always supported my work and inspired me to be a better researcher and a better teacher. Also thank you to my mentor in teaching, textbooks and computing, the late Hans Petter Langtangen. Without you, this book would never have been realized. Your vision, voice and spirit live on in us who worked with you. And thank you to my colleague Morten Hjorth-Jensen who has built up the group in computational physics at the University of Oslo, who generously included me in this group, and who has by example inspired me to be a better teacher.

This textbook was written using doconce—a document translation and formatting tool that allows simple integration of text, mathematics and computer code developed by Hans Petter Langtangen.

Oslo, Norway   Anders Malthe-Sørenssen
February 2024

Contents

Introduction to Percolation

1

In this chapter we motivate the study of disordered media through the example of a porous system. The basic terms in percolation theory are introduced, and you learn how to generate, visualize and measure on percolation systems in Python. We demonstrate how to find exact results for small systems in two dimensions by addressing all configurations of the system, and show that this approach becomes unfeasible for large system sizes.

Percolation is the study of connectivity of random media and of other properties of connected subsets of random media [8, 30, 37]. Figure 1.1 illustrates a porous material—a material with holes, pores, of various sizes. This is an example of a random material with built-in disorder. In this book, we will address the physical properties of such media, develop the underlying mathematical theory and the computational and statistical methods needed to discuss the physical properties of random media. In order to do that, we will develop a simplified model system, a model porous medium, for which we can develop a well-founded mathematical theory, and then afterwards we can apply this model to realistic random systems.

A Porous Medium as a Model of a Disordered Material The porous medium illustrated in the figure serves as a useful, fundamental model for random media in general. What characterizes the porous material in Fig. 1.1? The porous material consists of regions with and without material. It is therefore an extreme, binary version of a random medium. An actual physical porous material will be generated by some physical process, which will affect the properties of the porous medium in some way. For example, if the material is generated by sedimentary deposition, details of the deposition process may affect the shape and connectivity of the pores, or later fracturing may generate straight, open cracks in addition to more round pores. These features are always present in the complex geometries found in nature, and they will generate correlations in the randomness of the material. While these correlations can be addressed in detailed, specific studies of random materials, we

A. Malthe-Sørenssen, *Percolation Theory Using Python*, Lecture Notes in Physics 1029, https://doi.org/10.1007/978-3-031-59900-2_1

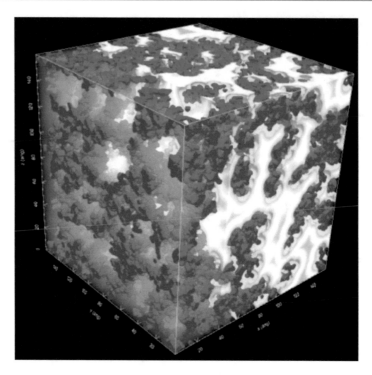

Fig. 1.1 Illustration of a porous material from a simulation of nanoporous silicate (SiO_2). The colors inside the pores illustrates the distance to the nearest part of the solid

will here instead start with a simpler class of materials—*uncorrelated random, porous materials.*

A Simplified Model of a Porous Medium We will here introduce a simplified model for a random porous material. We divide the material into cubes (3d) or squares (2d), called *sites*, of size d. Each site can be either filled or empty. We can use this method to characterize an actual porous medium, as illustrated in Fig. 1.1, or we can use it as a model for a *random porous medium* if we fill each site with a probability p. On average, the volume of the solid part of the material will be $V_s = pV$, where V is the volume of the system, and the volume of the pores will be $V_p = (1 - p)V$. We usually call the relative volume of the pores, the **porosity**, $\phi = V_p/V$, of the material. The solid is called the **matrix** and the relative volume of the matrix, V_s/V is called the solid fraction, which is denoted by $c = V_s/V$. In this case, we see that p corresponds to the solid fraction. Initially, we will assume that on the scale of lattice cells, the probabilities for sites to be filled are statistically independent—we will study an *uncorrelated random medium.*

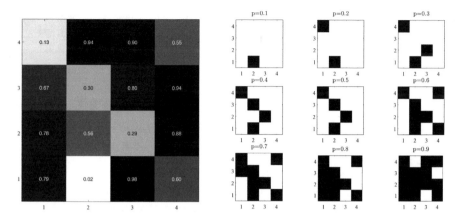

Fig. 1.2 Illustration of an array of 4×4 random numbers, and the set sites for different values of p

Generating a Random Medium in Python Figure 1.2 illustrates a two-dimensional system of 4×4 cells. The cells are filled with a probability p. We will call the filled cells occupied or set, and they are colored black. To generate such a matrix with first generate a matrix z with elements z_i that are uniform random numbers between 0 and 1. A given site i is set if $z_i \leq p$ and it is empty otherwise. This is implemented by

```
import numpy as np
import matplotlib.pyplot as plt
p = 0.25
z = np.random.rand(4,4)
m = z<p
plt.imshow(m)
```

The resulting matrices are shown in Fig. 1.2 for various values of p. The left figure illustrates the random values in the matrix, z, and the right figures the set sites for various values of p. You can think of this process as similar a to changing the water level in a landscape (z_i) and observing what parts of a landscape is below water ($z_i \leq p$).

Connectivity in a Random Medium Percolation is the study of connectivity. The simplest question we can ask is: When does a path form from one side of the sample to the other? By when, we mean at what value of p. For the particular realizations of the matrix m in Fig. 1.2 we see that the answer depends on how we define connectivity. If we want to make a path along the set (black) sites from one side to another, we must decide on when two sites are connected. Here, we will typically use *nearest neighbor connectivity*: Two sites in a square (cubic) lattice are connected if they are nearest neighbors. In the square lattice in Fig. 1.2, each site has $Z = 4$ nearest neighbors and $Z = 8$ next-nearest neighbors, where the number Z

is called the coordination number. We see that with nearest-neighbor connectivity, we get a path from the bottom to the top when $p = 0.7$, but with next-nearest neighbor connectivity we would get a path from the bottom to the top already at $p = 0.4$. We call the value p_c, the lowest value of p where we get a connected path from one side to another (from the top to the bottom, from the left to the right, or both) the *percolation threshold*. For a given realization of the matrix, there is well-defined value for p_c, but another realization would give another realization of p_c. We therefore need to either use statistical averages to characterize the properties of the percolation system, or we need to refer to a theoretical—thermodynamic—limit, such as the value for p_c in an infinitely large system. When we use p_c here, we will usually refer to the thermodynamic value.

In this book, we will develop theories describing various physical properties of the percolation system as a function of p. We will characterize the sizes of connected regions, the size of the region connecting one side to another, the size of the region that contributes to transport (fluid, thermal or electrical transport), and other geometrical properties of the system. Most of the features we study will be universal, that is, independent of many of the details of the system. From Fig. 1.2 we see that p_c depends on the details. For example, it depends on the definition of connectivity. It would also depend on the type of lattice used: square, triangular, hexagonal, etc. The value of p_c is specific. However, many other properties of the system are general. For example, how the conductivity of the porous medium depends on p near p_c does not depend on the type of lattice or the choice of connectivity rule. It is universal. This means that we can choose a system which is simple to study in order to gain intuition about the general features, and then apply that intuition to the special cases afterwards.

While the connectivity or type of lattice does not matter, some things do matter. For example, the dimensionality matters: The behavior of a percolation system is different in one, two and three dimensions. However, the most important changes in behavior occur between one and two dimensions, where the difference is dramatic, whereas the difference between two and three dimensions is more of a degree that we can easily handle. Actually, the percolation problem becomes simpler again in higher dimensions. In two dimensions, it is possible for a path to around a hole and still have connectivity. But it is not possible to have connectivity of both the pores and the solid in the same direction at the same time. This is possible in three dimensions: A two-dimensional creature would have problems with having a digestive tract, as it would divide the creature in two, but in three dimensions this is fully possible. Here, we will therefore focus on two and three-dimensional systems.

We will first address percolation in one and infinite dimensions, since we can solve the problems exactly in these cases. We will then address percolation in two dimensions, where there are no exact solutions. However, we will see that if we assume that the distribution of cluster sizes has a particular scaling form, we can still address the problem in two dimensions and make powerful predictions. We will also see that close to the percolation threshold the porous medium has a self-affine scaling structure—it is a fractal. This property has important consequences for the physical properties of random systems. We will also see how this is reflected in a systematic change of scales, a renormalization procedure, which is a general tool that can applied to rescaling in many areas.

1.1   Basic Concepts in Percolation

Let us start by studying a specific example of a random medium. We will generate an $L \times L$ lattice of points, called sites, that are occupied with probability p. This corresponds to a coarse-grained porous medium with a porosity $\phi = p$, if we assume that the occupied sites are holes in the porous material and look at the connectivity of the pores in the material.

We can generate a realization of a square $L \times L$ system in python using

```
import numpy as np
import matplotlib.pyplot as plt
L = 20
p = 0.5
z = np.random.rand(L,L)
m = z<p
plt.imshow(m, origin='lower')
```

The resulting matrix is illustrated in Fig. 1.3. However, this visualization does not provide us with any insight into the connectivity of the sites in this system. Let us instead analyze the connected regions in the system.

Definitions

- two sites are **connected** if they are nearest neighbors (there are 4 nearest neighbors on a square lattice)
- a **cluster** is a set of connected sites
- a cluster is **spanning** if it spans from one side to the opposite side
- a cluster that is spanning is called the **spanning cluster**
- a system is **percolating** if there is a spanning cluster in the system

---

Fortunately, there are built-in functions in python that finds connected regions in an image.[1] The function `measurements.label` finds clusters based on a given connectivity. For example, with a coordination number $Z = 4$, that is nearest neighbor connectivity, we find

```
from scipy.ndimage import measurements
lw, num = measurements.label(m)
```

---

[1] Notice that these program lines will need the libraries numpy, matplotlib.pyplot and measurements to be loaded. We will assume you have done this in the following.

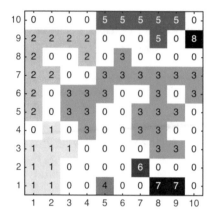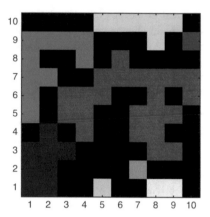

Fig. 1.3 Illustration of the index array for a 10×10 system for $p = 0.45$

This function returns the matrix `lw`, which for each site in the original array tells what cluster it belongs to. Clusters are numbered sequentially, and each cluster is given an index number. All the sites with the same index number belong to the same cluster. The resulting array is shown in Fig. 1.3, where the index number for each site is shown and a color is used to indicate the clusters. Notice that there is a distribution of cluster sizes, but no cluster is large enough to reach from one side to another, and as a result the system does not percolate.

In order to visualize the clusters effectively, we give the various clusters different colors.

```
plt.imshow(lw, origin='lower')
```

Unfortunately, this colors the clusters gradually from the bottom up. This is a property of the underlying algorithm: Clusters are indexed starting from the top-left of the matrix (which is the bottom-left of the image). Hence, clusters that are close to each other will get similar colors and can therefore be difficult to discern unless we shuffle the colormap. We can fix this by shuffling the labeling:

```
b = np.arange(lw.max() + 1)
np.random.shuffle(b)
shuffledLw = b[lw]
plt.imshow(shuffledLw, origin='lower')
```

The resulting image is shown to the right in Fig. 1.3. (Notice that in these figures we have reversed the ordering of the y-axis. Usually, the first row is in the top-left corner in your plots, but when we use the keyword `lower` the first row is in the bottom-left).

It may also be useful to color the clusters based on the size of the clusters, where size refers to the number of sites in a cluster. We can do this using

```
area = measurements.sum(m, lw, index=np.arange(lw.max() + 1))
areaImg = area[lw]
```

```
plt.imshow(areaImg, origin='lower')
plt.colorbar()
```

Let us now study the effect of p on the set of connected clusters. We vary the value of p for the same underlying random matrix, and plot the resulting images:

```
import numpy as np
import matplotlib.pylab as plt
from scipy.ndimage import measurements
plt.figure(figsize=(10,8))
L = 100
pv = [0.2,0.3,0.4,0.5,0.6,0.7]
z = np.random.rand(L,L)
for i in range(len(pv)):
    p = pv[i]
    m = z<p
    lw, num = measurements.label(m)
    area = measurements.sum(m, lw, index=np.arange(lw.max() + 1))
    areaImg = area[lw]
    plt.subplot(2,3,i+1)
    tit = 'p='+str(p)
    plt.imshow(areaImg, origin='lower')
    plt.title(tit)
```

Figure 1.4 shows the clusters for a 100×100 system for p ranging from 0.2 to 0.7 in steps of 0.1. We see that the clusters increase in size as p increases. At $p = 0.6$ there is one large cluster spanning the entire region. We have a *percolating cluster*, and we call the cluster that spans the system the **spanning cluster**. The transition is very rapid from $p = 0.5$ to $p = 0.6$. We therefore look at this region in more detail

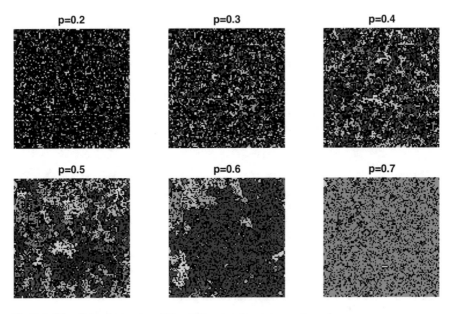

Fig. 1.4 Plot of the clusters in a 100×100 system for various values of p

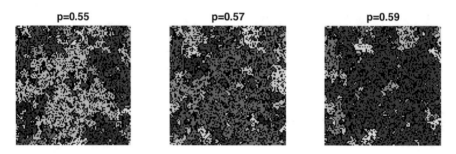

Fig. 1.5 Plot of the clusters in a 100×100 system for various values of p

in Fig. 1.5. We see that the size of the largest cluster increases rapidly as p reaches a value around 0.6, which corresponds to p_c for this system. At this point, the largest cluster spans the entire system. For the two-dimensional system illustrated here we know that in an infinite lattice the percolation threshold is $p_c \simeq 0.5927$.

The aim of this book is to develop a theory to describe how this random porous medium behaves close to p_c. We will characterize properties such as the density of the spanning cluster, the geometry of the spanning cluster, and the conductivity and elastic properties of the spanning cluster. We will address the distribution of cluster sizes and how various parts of the clusters are important for particular physical processes. We start by characterizing the behavior of the spanning cluster near p_c.

1.2   Percolation Probability

When does the system percolate? When there exists a path connecting one side to another. This occurs at some value $p = p_c$. However, in a finite system, like the system we simulated above, the value we find for p_c will vary with each realization. It may be slightly above or slightly below the p_c we find in an infinite sample. Later, we will develop a theory to understand how the effective p_c in a finite system varies from the thermodynamic p_c. But already now we realize that as we perform different numerical experiments, we will measure various values of p_c. We can characterize this behavior by introducing a probability $\Pi(p, L)$:

> **Percolation Probability** The percolation probability $\Pi(p, L)$ is the probability for there to be a connected path from one side to another side as a function of p in a system of size L.

We can measure $\Pi(p, L)$ in a finite sample of size $L \times L$, by generating many random matrices. For each matrix, we perform a cluster analysis for a sequence of p_i values. For each p_i we find all the clusters. We then check if any of the clusters are present both on the left and on the right side of the lattice. In that case, they are spanning (We could also have included a test for clusters spanning from the top to the bottom, but this does not change the statistics significantly). In this case, there is a spanning cluster—the system percolates. We count how many times, N_i, a system percolates for a given p_i and then divide by the total number of experiment, N, to estimate the probability for percolation for a given p_i, $\Pi(p_i, L) \simeq N_i/N$. We implement this as follows. First, we generate a sequence of 100 p_i values from 0.35 to 1.0:

```
p = np.linspace(0.35,1.0,100)
```

Then we prepare an array for N_i with the same number of elements as p_i:

```
nx = len(p)
Pi = np.zeros(nx)
```

We will generate $N = 1000$ samples:

```
N = 1000
```

We will then loop over all samples, and for each sample we generate a new random matrix. The for each value of p_i we perform the cluster analysis as we did above. We use the function measurements.label to label the clusters. Then we find the intersection between the labels on the left and the right side of the system and store in perc_x. If the length of the set of intersections is larger than zero, there is at least one percolating cluster, and we find the label of the spanning cluster(s) in perc:

```
lw,num = measurements.label(z)
perc_x = np.intersect1d(lw[0,:],lw[-1,:})
perc = perc_x[np.where(perc_x>0)]
```

Now, we are ready to implement this into a complete program. For a given value of p, we count in how many simulations $N_p(p)$ there is a path spanning from one side to another and estimate $\bar{\Pi}(p) \simeq N_p(p)/N$, where N is the total number of simulations/samples. This is implemented in the following program:

```
import numpy as np
import matplotlib.pyplot as plt
from scipy.ndimage import measurements
p = np.linspace(0.4,1.0,100)
nx = len(p)
Ni = np.zeros(nx)
P = np.zeros(nx)
N = 1000
L = 100
for i in range(N):
    z = np.random.rand(L,L)
    for ip in range(nx):
        m = z<p[ip]
        lw, num = measurements.label(m)
        perc_x = np.intersect1d(lw[0,:],lw[-1,:])
```

```
                perc = perc_x[np.where(perc_x>0)]
                if (len(perc)>0):
                    Ni[ip] = Ni[ip] + 1
    Pi = Ni/N
    plt.plot(p,Pi)
    plt.xlabel('$p$')
    plt.ylabel('$\Pi$')
```

The resulting plot of $\Pi(p, L)$ is seen in Fig. 1.6. The figure shows the resulting plots as a function of system size L. We see that as the system size increases, $\Pi(p, L)$ approaches a step function at $p = p_c$.

1.3 Spanning Cluster

The probability $\Pi(p, L)$ describes the probability for there to be a spanning cluster, but what about the spanning cluster itself, how can we characterize it? We see from Fig. 1.4 that the spanning cluster grows quickly around $p = p_c$. Let us therefore characterize the cluster by its size, M_S, or by its density, $P(p, L) = M_S/L^2$, which corresponds to the probability for a site to belong the spanning cluster.

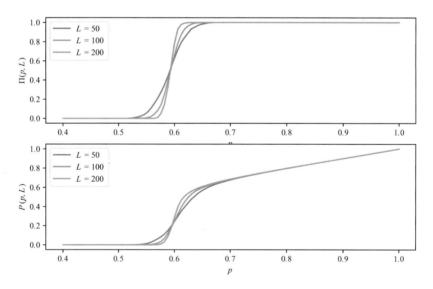

Fig. 1.6 Plot of $\Pi(p, L)$, the probability for there to be a connected path from one side to anther, as a function of p for various system sizes L, and $P(p, L)$, the density of the spanning cluster

Density of the Spanning Cluster The probability $P(p, L)$ for a site to belong to a spanning cluster is called the *density of the spanning cluster*.

We can measure $P(p, L)$ by counting the mass M_i of the spanning cluster as a function of p_i for various values of p_i. We can find the mass of the spanning cluster by finding a cluster that spans the system (there may be more than one) as we did above, and then measure the number of sites in the cluster using `area = measurements.sum(m, lw, perc)`.

We do this in the same program as we developed above. For each p_i, we see if a cluster is spanning from one side to another, and if it is, we add the mass of this cluster to $M_S(p_i)$. We implement these features in the following program, which measures both $\Pi(p, L)$ and $P(p, L)$ for a given value of L:

```python
import numpy as np
import matplotlib.pyplot as plt
from scipy.ndimage import measurements
p = np.linspace(0.4,1.0,100)
nx = len(p)
Ni = np.zeros(nx)
P = np.zeros(nx)
N = 1000
L = 100
for i in range(N):
    z = np.random.rand(L,L)
    for ip in range(nx):
        m = z<p[ip]
        lw, num = measurements.label(m)
        perc_x = np.intersect1d(lw[0,:],lw[-1,:])
        perc = perc_x[np.where(perc_x>0)]
        if (len(perc)>0):
            Ni[ip] = Ni[ip] + 1
            area = measurements.sum(m, lw, perc[0])
            P[ip] = P[ip] + area
Pi = Ni/N
P = P/(N*L*L)
plt.subplot(2,1,1)
plt.plot(p,Pi)
plt.ylabel("$\Pi(p)$")
plt.subplot(2,1,2)
plt.plot(p,P)
plt.ylabel("P(p)")
plt.xlabel("p")
```

The resulting plot of $P(p, L)$ is shown in the bottom of Fig. 1.6. We see that $P(p, L)$ changes rapidly around $p = p_c$, and that it grows slowly—approximately linearly—as $p \to 1$. We can understand this linear behavior: When p is near 1 practically all the set sites are connected and are part of the spanning cluster. In this limit, the density of the spanning cluster is therefore proportional to the number of sites that are present, which again is proportional to p. We will now develop a theory for the observations of $\Pi(p, L)$, $P(p, L)$ and other features of the percolation system. First, we see what insights we can gain from small, finite systems.

1.4   Percolation in Small Systems

First, we will address the two-dimensional system directly. We will study a $L \times L$ system, and the various physical properties of it. We will start with $L = 1$ and $L = 2$ and then try to generalize.

$L = 1$ First, let us address $L = 1$. In this case, the system percolates if the site is present, which has a probability p. The percolation probability is therefore $\Pi(p, 1) = p$. Similarly, the probability for a site to belong to the spanning cluster is p and therefore $P(p, 1) = p$.

$L = 2$ Then, let us examine $L = 2$. This is still simple, but we now have to develop a more advanced strategy than for $L = 1$. Our strategy will be to list all possible outcomes, find the probability for each outcome, and then use this to find the probability for the various physical properties we are interested in. The possible configurations are shown in Fig. 1.7.

Our plan is to use a basic result from probability theory: If we want to calculate the probability of an event A, we can do this by summing the probability of A given B multiplied by the probability for B over all possible outcomes B (as long as the set of outcomes B span the space of all outcomes and are mutually exclusive, that is, that they have no intersection). In this case:

$$P(A) = \sum_B P(A|B)P(B) \,, \qquad (1.1)$$

where we have used the notation $P(A|B)$ to denote the conditional probability of A given that B occurs. We can use this to calculate properties such as $\Pi(p, L)$ and $P(p, L)$ by summing over all possible configurations c of the system:

$$\Pi(p, L) = \sum_c \Pi(p, L|c)P(c) \,, \qquad (1.2)$$

where $\Pi(p, L|c)$ is the value of Π for the particular configuration c, and $P(c)$ is the probability of this configuration.

The configurations for $L = 2$ have been numbered from $c = 1$ to $c = 16$ in Fig. 1.7. However, configurations that are either mirror images or rotations of each other will have the same probability and the same physical properties since percolation can take place both in the x and the y directions. It is therefore only necessary to group the configurations into 6 different classes, k, as illustrated in the bottom of Fig. 1.7, but we then need to include the multiplicity, g_k, for each class

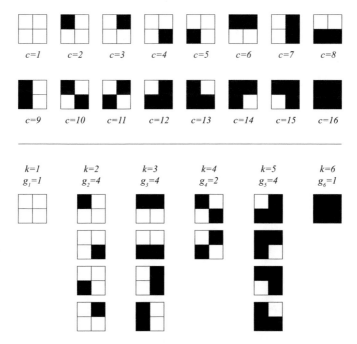

Fig. 1.7 The possible configurations for a $L = 2$ site percolation lattice in two-dimensions. The configurations are indexed using the cluster configuration number c

when we calculate probabilities. The probability $\Pi(p, L)$ is then:

$$\Pi(p, L) = \sum_k g_k \Pi(p, L|k) P(k) . \tag{1.3}$$

Table 1.1 lists the classes, the number of configurations in each class, the probability of *one* such configuration in a class, and the value of $\Pi(p, L|k)$ for this class.

We should check that we have actually listed all possible configurations. In general, the number of configurations for an $L \times L$ system is 2^{L^2}. The total number of configurations is $1 + 4 + 2 + 4 + 4 + 1 = 16$, which is equal to 2^4 as it should. We have therefore included all the configurations.

Table 1.1 List of classes, configurations in each class, and the probability of one such configuration

c	g_k	$P(k)$	$\Pi(p, L|k)$
1	1	$p^0(1-p)^4$	0
2	4	$p^1(1-p)^3$	0
3	4	$p^2(1-p)^2$	1
4	2	$p^2(1-p)^2$	0
5	4	$p^3(1-p)^1$	1
6	1	$p^4(1-p)^0$	1

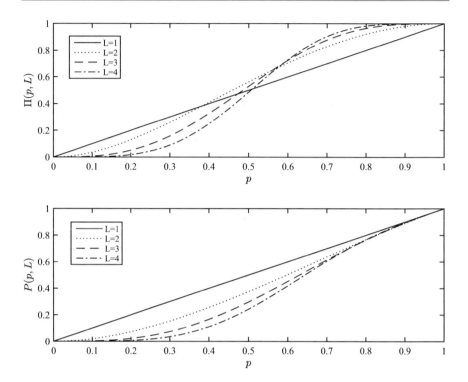

Fig. 1.8 Plot of $\Pi(p, L)$ for $L = 1$ and $L = 2$ as a function of p

We can then find the probability for Π by direct calculation of the sum:

$$\Pi = 0 \cdot 1 \cdot p^0 (1 - p)^4 + 0 \cdot 4 \cdot p^1 (1 - p)^3 + 1 \cdot 4 \cdot p^2 (1 - p)^2 \tag{1.4}$$

$$+ 0 \cdot 2 \cdot p^2 (1 - p)^2 + 1 \cdot 4 \cdot p^3 (1 - p)^1 + 1 \cdot 1 \cdot p^4 (1 - p)^0 . \tag{1.5}$$

The exact value for $\Pi(p, L = 2)$ is therefore:

$$\Pi(p, L = 2) = 4p^2 (1 - p)^2 + 4p^3 (1 - p)^1 + p^4 (1 - p)^0 , \tag{1.6}$$

which we can simplify further if we want. The shape of $\Pi(p, L)$ for $L = 1$, and $L = 2$ is shown in Fig. 1.8.

Estimating p_c We could characterize $p = p_c$ as the number for which $\Pi(p_c) = 1/2$. For $L = 1$, we then get $\Pi(p_c) = p_c = 1/2$. And for $L = 2$, we find $4p_c^2(1 - p_c)^2 + 4p_c^3(1 - p_c)^1 + p_c^4(1 - p_c)^0 = 1/2$, which gives $p_c(L = 2) \simeq 0.4588$. Maybe we can just continue doing this type of calculation for higher and higher L and we will get a better and better approximation for p_c?

Extending to Larger Systems We notice that for finite L, $\Pi(p, L)$ will be a polynomial of order $\mathcal{O}(L^2)$ - it is in principle a function we can calculate. However, the number of possible configurations is 2^{L^2} which increases very rapidly with L. It is therefore not realistic to use this technique for calculating the percolation probabilities. We will need to have more powerful techniques, or simpler problems, in order to perform exact calculations.

However, we can still learn much from a discussion of finite L. For example, we notice that

$$\Pi(p, L) \simeq Lp^L + c_1 p^{L+1} + \ldots + c_n p^{L^2} , \tag{1.7}$$

in the limit of $p \ll 1$. The leading order term when $p \to 0$ is therefore Lp^L.

Similarly, we find that for $p \to 1$, the leading order term is approximately

$$\Pi(p, L) \simeq 1 - (1 - p)^L . \tag{1.8}$$

These two results gives us an indication about how the percolation probability $\Pi(p, L)$ is approaching the step function when $L \to \infty$.

Similarly, we can calculate $P(p, L)$ for $L = 2$. However, we leave the calculation of the $L = 3$ and the $P(p, L)$ system to the exercises.

1.5 Further Reading

There are good general introduction texts to percolation theory such as the popular books by Stauffer and Aharony [37], by Sahimi [30], by Christensen and Moloney [8], and the classical book by Grimmet [14]. Mathematical aspects are addressed by Kesten [21] and phase transitions in general are introduced by e.g. Stanley [35]. Applications of percolation theory are found in many fields such as in geoscience [22], porous media [18] or social networks [33] and many more. We encourage you to explore these books for a more theoretical introduction to percolation theory.

Exercises

Exercise 1.1 (Percolation for $L = 3$)

(a) Find $P(p, L)$ for $L = 1$ and $L = 2$.
(b) Categorize all possible configurations for $L = 3$.
(c) Find $\Pi(p, L)$ and $P(p, L)$ for $L = 3$.

Exercise 1.2 (Counting Configurations in Small Systems)

(a) Write a program to find all the configurations for $L = 2$.
(b) Use this program to find $\Pi(p, L = 2)$ and $P(p, L = 2)$. Compare with the exact results from the previous exercise.
(c) Use you program to find $\Pi(p, L)$ and $P(p, L)$ for $L = 3, 4$ and 5.

Exercise 1.3 (Percolation in Small Systems in 3d)   In this exercise we will study the three-dimensional site percolation system for small system sizes.

(a) How many configurations are there for $L = 2$?
(b) Categorize all possible configurations for $L = 2$.
(c) Find $\Pi(p, L)$ and $P(p, L)$ for $L = 2$.
(d) Compare your results with your result for the two-dimensional system. Comment on similarities and differences.

One-Dimensional Percolation

2

The percolation problem can be solved exactly in two limits: in the one-dimensional and the infinite dimensional cases. Here, we will first address the one-dimensional system. While the one-dimensional system does not allow us to study the full complexity of the percolation problem, many of the concepts and measures introduced to study the one-dimensional problem can generalized to higher dimensions.

2.1 Percolation Probability

Let us first address percolation in a one-dimensional lattice of L sites. In this case, there is a spanning cluster if and only if all the sites are occupied. If only a single site is empty, the connecting path will be broken and there will not be any connecting path from one side to the other. The percolation probability is therefore

$$\Pi(p, L) = p^L . \tag{2.1}$$

This has a trivial behavior when $L \to \infty$

$$\Pi(p, \infty) = \begin{cases} 0 & \text{when } p < 1 \\ 1 & \text{when } p = 1 \end{cases} . \tag{2.2}$$

This shows that the percolation threshold is $p_c = 1$ in one dimension. However, the one-dimensional system is anomalous, and in higher dimensions we will always have $p_c < 1$, so that we can study the system both above and below p_c. Unfortunately, for the one-dimensional system we can only study the system below p_c.

© The Author(s) 2024
A. Malthe-Sørenssen, *Percolation Theory Using Python*, Lecture Notes
in Physics 1029, https://doi.org/10.1007/978-3-031-59900-2_2

2.2   Cluster Number Density

Definition of Cluster Number Density

In the simulations in Fig. 1.4 we saw that the percolation system was characterized by a wide distribution of clusters—regions of connected sites. The clusters have varying shape and size. If we increase p to approach p_c we saw that the clusters increased in size until they reached the system size. We can use the one-dimensional system to learn more about the behavior of clusters as p approaches p_c.

Figure 2.1 illustrates a realization of an $L = 16$ percolation system in one dimension below $p_c = 1$. In this case there are 5 clusters of sizes: 1,1,4,2,1 measured as the number of sites in each cluster. The clusters are numbered, indexed, from 1 to 5 as we did for the numerical simulations in two dimensions. How can we characterize the clusters in a system? In percolation theory we characterize cluster sizes by asking a particular question: If you point at a (random) site in the lattice, what is the probability for this site to belong to a cluster of size s?

$$P(\text{site is part of cluster of size } s) = sn(s, p) . \tag{2.3}$$

It is common to use the notation $sn(s, p)$ for this probability for a given site to belong to a cluster of size s. Why is it divided into two parts, s and $n(s, p)$? Because we must divide the question into two parts: (1) What is the probability for a given site to be a *specific site* in a cluster of size s, and (2) how many such specific sites are there in a cluster? What do we mean by a specific site? For the cluster with index 3 in Fig. 2.1 there are 4 sites. We could therefore ask the question: What is the probability for a site to be the left-most site in a cluster of size s? This is what

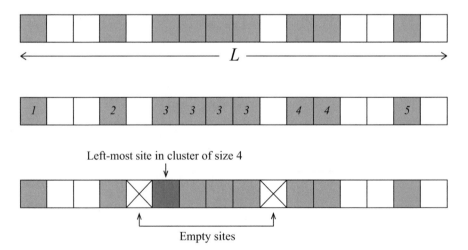

Fig. 2.1 Realization of a $L = 16$ percolation system in one dimension. Separate clusters are illustrated by the indexes, shown as numbers inside the sites. Occupied sites are marked with gray squares

we mean with a specific site. We could ask the same question about the second left-most, the third left-most and so on. We call the probability for a site to belong to a specific site in a cluster of size s (such as the left-most site in the cluster) the **cluster number density**, and we use the notation $n(s, p)$ for this. To find the probability $sn(s, p)$ for a site to belong to any of the s sites in a cluster of size s we must sum the probabilities for each of the specific sites. This is illustrated for the case of a cluster of size 4:

$$P(\text{site to be in cluster of size 4})$$

$$= P(\text{site to be left-most site in cluster of size 4})$$

$$+ P(\text{site to be second left-most site in cluster of size 4})$$

$$+ P(\text{site to be third left-most site in cluster of size 4})$$

$$+ P(\text{site to be fourth left-most site in cluster of size 4})$$

$$= 4P(\text{site to be left-most site in cluster of size 4}) ,$$

because each of these probabilities are the same. What is the probability for a site to be the left-most site in a cluster of size s in one dimension? In order for it to be in a cluster of size s, the site must be present, which has probability p, and then $s - 1$ sites must also be present to the right of it, which has probability p^{s-1}. In addition, the site to the left must be empty (illustrated by an X in Fig. 2.1 bottom part), which has probability $(1 - p)$ and the site to the right of the fourth site (illustrated by an X in Fig. 2.1 bottom part), which also has probability $(1 - p)$. Since the occupation probabilities for each site are independent, the probability for the site to be the left-most site in a cluster of size s is the product of these probabilities:

$$n(s, p) = p\,p^{s-1}\,(1 - p)\,(1 - p) = (1 - p)^2 p^s . \tag{2.4}$$

This is the cluster number density in one dimension.

> **Cluster Number Density** The cluster number density $n(s, p)$ is the probability for a site to be a particular site in a cluster of size s. For example, in one dimension, $n(s, p)$ can be interpreted as the probability for a site to be the left-most site in a cluster of size s.

We should check that $sn(s, p)$ really is a normalized probability. How should it be normalized? We know that if we point at a random site in the system, the probability for that site to be occupied is p. An occupied site is then either a part of a finite cluster of some size s or it is part of the infinite cluster. The probability for a site to be a part of the infinite cluster we called P. This means that we have the following normalization condition:

Normalization of the Cluster Number Density A site is occupied with probability p. An occupied site is either part of a finite cluster of size s with probability $sn(s, p)$ or it is part of the infinite (spanning) cluster with probability $P(p)$:

$$p = \sum_{s=1}^{\infty} sn(s, p) + P(p) . \tag{2.5}$$

Let us check that this is indeed the case for the one-dimensional expression for $n(s, p)$ by calculating the sum:

$$\sum_{s=1}^{\infty} sn(s, p) = \sum_{s=1}^{\infty} sp^s(1 - p)^2 = (1 - p)^2 p \sum_{s=1}^{\infty} sp^{s-1} , \tag{2.6}$$

where we will now employ a common trick:

$$\sum_{s=1}^{\infty} sp^{s-1} = \frac{d}{dp} \sum_{s=0}^{\infty} p^s = \frac{d}{dp} \frac{1}{1 - p} = (1 - p)^{-2} , \tag{2.7}$$

which gives

$$\sum_{s=1}^{\infty} sn(s, p) = (1 - p)^2 \, p \sum_{s=1}^{\infty} sp^{s-1} = (1 - p)^2 \, p \, (1 - p)^{-2} = p . \tag{2.8}$$

Since $P = 0$ when $p < 0$ we see that the probability is normalized. We can use similar tricks to calculate moments of any order.

Measuring the Cluster Number Density

In order to gain further insight into the distribution of cluster sizes, let us study Fig. 2.1 in more detail. There are 3 clusters of size $s = 1$, one cluster of size $s = 2$, and one cluster of size $s = 4$. We could therefore introduce a histogram of cluster sizes, which is what we would do if we studied the cluster distribution numerically. Let us write N_s as the number of clusters of size s so that $N_1 = 3$, $N_2 = 1$, $N_3 = 0$ and $N_4 = 1$.

How can we estimate $sn(s, p)$, the probability for a given site to be part of a cluster of size s, from N_s? The probability for a site to belong to cluster of size s can be estimated by the number of sites belonging to a cluster of size s divided by the total number of sites. The number of sites belonging to a cluster of size s

is sN_s, and the total number of sites is L^d, where L is the system size and d is the dimensionality. (Here, $d = 1$). This means that we can estimate the probability $sn(s, p)$ from

$$\overline{sn(s, p)} = \frac{sN_s}{L^d} , \qquad (2.9)$$

where we use a bar to show that this is an estimated quantity and not the actual probability. We divide by s on both sides, and find

$$\overline{n(s, p)} = \frac{N_s}{L^d} . \qquad (2.10)$$

This argument and the result are valid in any dimension, not only for $d = 1$. We can also see why this quantity is called the cluster number density: it is the number of clusters divided by the volume measured in number of sites. We have therefore found a method to estimate the cluster number density:

Measuring the Cluster Number Density We can measure $n(s, p)$ in a simulation by measuring N_s, the number of clusters of size s, and then calculate $n(s, p)$ from

$$\overline{n(s, p)} = \frac{N_s}{L^d} . \qquad (2.11)$$

For the clusters in Fig. 2.1 we find that

$$\overline{n(1, p)} = \frac{N_1}{L^1} = \frac{3}{16} , \qquad (2.12)$$

$$\overline{n(2, p)} = \frac{N_2}{L^1} = \frac{1}{16} , \qquad (2.13)$$

$$\overline{n(3, p)} = \frac{N_3}{L^1} = \frac{0}{16} , \qquad (2.14)$$

$$\overline{n(4, p)} = \frac{N_4}{L^1} = \frac{1}{16} , \qquad (2.15)$$

which is our estimate of $n(s, p)$ based on this single realization. We check the consistency of the result by ensuring that the estimated probabilities also are normalized:

$$\sum_s \overline{sn(s, p)} = 1 \cdot \frac{3}{16} + 2 \cdot \frac{1}{16} + 3 \cdot 0 + 4 \cdot \frac{1}{16} = \frac{9}{16} = \overline{p} , \qquad (2.16)$$

where $\overline{p}$ is estimated from the number of present sites divided by the total number of sites.

In order to produce good statistical estimates for $n(s, p)$, we must sample from many random realization of the system. If we sample from M realizations, and then measure the total number of clusters of size s, $N_s(M)$, summed over all the realizations, we estimate the cluster number density from

$$\overline{n(s, p)} = \frac{N_s(M)}{ML^d} . \qquad (2.17)$$

Notice that all simulations are for finite L, and we would therefore expect deviations due to a finite L as well as due to the finite number of samples. However, we expect the estimated $\overline{n(s, p; L)}$ to approach the underlying $n(s, p)$ as M and L approaches infinity.

Shape of the Cluster Number Density

We found that the cluster number density in one dimension is

$$n(s, p) = (1 - p)^2 p^s . \qquad (2.18)$$

In Fig. 2.2 we have plotted $n(s, p)$ for various values of p. In order to compare the s-dependence of the plot directly for various p-values we plot

$$G(s) = (1 - p)^2 n(s, p) = p^s , \qquad (2.19)$$

as a function of s. We notice that $(1 - p)^2 n(s, p)$ is approximately constant for a wide range of s and then falls off rapidly for some characteristic value s_ξ which increases as p approaches $p_c = 1$. We can understand this behavior better by rewriting $n(s, p)$ as

$$n(s, p) = (1 - p)^2 e^{s \ln p} = (1 - p)^2 e^{-s/s_\xi} , \qquad (2.20)$$

where we have introduced the cut-off cluster size

$$s_\xi = -\frac{1}{\ln p} . \qquad (2.21)$$

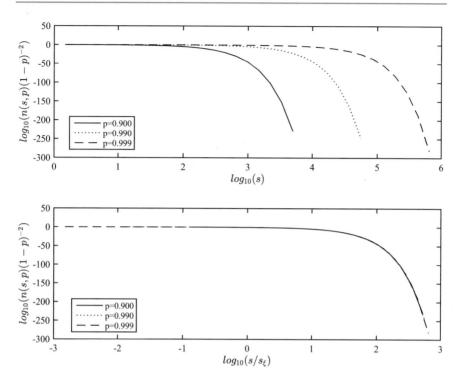

Fig. 2.2 (Top) A plot of $n(s, p)(1 - p)^2$ as a function of s for various values of p for a one-dimensional percolation system shows that the cut-off increases as a function of s. (Bottom) When the s axis is rescaled by s_ξ to s/s_ξ, all the curves fall onto a common scaling function, that is, $n(s, p) = (1 - p)^2 F(s/s_\xi)$

What we are seeing in Fig. 2.2 is therefore the exponential cut-off curve, where the cut-off $s_\xi(p)$ increases as $p \to 1$. We call it a *cut-off* because the value of $n(s, p)$ decays very rapidly (exponentially fast) when s is larger than s_ξ.

How does s_ξ depend on p?. We see from (2.21) that as p approaches $p_c = 1$, the characteristic cluster size s_ξ will diverge. The form of the divergence can be determined in more detail through a Taylor expansion:

$$s_\xi = -\frac{1}{\ln p} \tag{2.22}$$

when p is close to 1, we see that $1 - p \ll 1$ and we can write

$$\ln p = \ln(1 - (1 - p)) \simeq -(1 - p) , \tag{2.23}$$

where we have used that $\ln(1 - x) = -x + \mathcal{O}(x^2)$, which is simply the Taylor expansion of the logarithm, where $\mathcal{O}(x^2)$ is term that is on the order of x^2. As a result

$$s_\xi \simeq \frac{1}{1 - p} = \frac{1}{p_c - p} = |p - p_c|^{-1} . \tag{2.24}$$

This shows that the divergence of s_ξ as p approaches p_c is a power-law with exponent -1. This power-law behavior is general in percolation theory:

Scaling Behavior of the Characteristic Cluster Size The characteristic cluster size s_ξ diverges as

$$s_\xi \propto |p - p_c|^{-1/\sigma} , \tag{2.25}$$

when $p \to p_c$. In one dimension, $\sigma = 1$.

The value of the exponent σ depends on the lattice dimensionality, but it does not depend on the details of the lattice. It would, for example, be the same also for next-nearest neighbor connectivity.

The functional form we have found is also an example of a **data collapse**. We see that if we plot $(1 - p)^{-2} n(s, p)$ as a function of s/s_ξ, all data-points for various values of p should fall onto a single curve:

$$n(s, p) = (1 - p)^2 e^{-s/s_\xi} \Rightarrow (1 - p)^{-2} n(s, p) = e^{-s/s_\xi} , \tag{2.26}$$

as illustrated in Fig. 2.2. We call this a data-collapse. We have one behavior for small s and then a rapid cut-off when s reaches s_ξ. We can rewrite $n(s, p)$ so that all the s_ξ dependence is in the cut-off function by realizing that since $s_\xi \simeq (1 - p)^{-1}$ we have that $(1 - p)^2 = s_\xi^{-2}$. This gives

$$n(s, p) = s_\xi^{-2} e^{-s/s_\xi} = s^{-2} \left(\frac{s}{s_\xi} \right)^2 e^{-s/s_\xi} = s^{-2} F \left(\frac{s}{s_\xi} \right) . \tag{2.27}$$

where $F(u) = u^2 e^{-u}$. We will see later that this form for $n(s, p)$ is general—it is valid for percolation in any dimension, although with other values for the exponent -2 and other shapes of the cut-off function $F(u)$. In percolation theory, we call this exponent τ:

$$n(s, p) = s^{-\tau} F(s/s_\xi) , \tag{2.28}$$

where $\tau = 2$ in two dimensions. The exponent τ is another example of a universal exponent that does not depend on details such as the connectivity rule, while it depends on the dimensionality of the system.

Numerical Measurement of the Cluster Number Density

Let us now test the measurement method and the theory through a numerical study of the cluster number density. According to the theory developed above we can estimate the cluster number density $n(s, p)$ from

$$\overline{n(s, p)} = \frac{N_s(M)}{L^2 M} \, , \tag{2.29}$$

where $N_s(M)$ is the number of clusters of size s measured in M realizations of the percolation system. We generate a one-dimensional percolation system and index the clusters using

```
import numpy as np
from scipy.ndimage import measurements
L = 20
p = 0.90
z = np.random.rand(L)
m = z<p
lw, num = measurements.label(m)
```

Now, `lw` contains the indices for all the clusters. We can extract the size of the clusters by summing the number of elements for each label:

```
labelList = np.arange(lw.max() + 1)
area = measurements.sum(m, lw, labelList)
```

The resulting list of areas for one sample is

```
>> lw
array([1, 1, 1, 0, 2, 2, 2, 2, 2, 2, 2, 2, 2, 0, 3, 0, 4, 4,
       4, 4], dtype=int32)
>> area
array([0., 3., 9., 1., 4.])
```

We need to collect all the areas of all the clusters for many realizations, and then calculate the number of clusters of each size s based on this long list of areas. This is all brought together by continuously appending the `area`-array to the end of an array `allarea` that contains the areas of all the clusters.

```
import numpy as np
import matplotlib.pyplot as plt
from scipy.ndimage import measurements
nsamp = 1000
L = 1000
p = 0.90
```

```
allarea = np.array([])
for i in range(nsamp):
    z = np.random.rand(L)
    m = z<p
    lw, num = measurements.label(m)
    labelList = np.arange(lw.max() + 1)
    area = measurements.sum(m, lw, labelList)
    allarea = np.append(allarea,area)
n,sbins = np.histogram(allarea,bins=int(max(allarea)))
s = 0.5*(sbins[1:]+sbins[:-1])
nsp = n/(L*nsamp)
sxi = -1.0/np.log(p)
nsptheory = (1-p)**2*np.exp(-s/sxi)
plt.plot(s,nsp,'o',s,nsptheory,'-')
plt.xlabel('$s$')
plt.ylabel('$n(s,p)$')
```

This script also calculates N_s using the histogram function with L bins to ensure that there is at least one bin for each value of s:

```
n,sbins = np.histogram(allarea,bins=int(max(allarea)))
s = 0.5*(sbins[1:]+sbins[:-1])
```

where we find s as the midpoints of the bins returned by the histogram-function. We estimate $\overline{n(s, p)}$ from

```
nsp = n/(L*nsamp)
```

For comparison with theory, we calculate values from the theoretically predicted expression $n(s, p)$, which is $n(s, p) = (1 - p)^2 \exp(-s/s_\xi)$, where $s_\xi = -1/\ln p$. This is calculated for the same values of s as used for the histogram using:

```
sxi = -1.0/np.log(p)
nsptheory = (1-p)**2*np.exp(-s/sxi)
```

When we use the histogram-function with many bins, we risk that many of the bins contain zero elements. To remove these elements from the plot, we can use the nonzero function from numpy to find the indices of the elements of n that are non-zero:

```
i = np.nonzero(n)
```

And then we only plot the values of $\overline{n(s, p)}$ at these indices. The values for the theoretical $n(s, p)$ are calculated for all values of s, and the two are plotted in the same plot:

```
plt.plot(s[i],nsp[i],'o',s,nsptheory,'-')
```

The resulting plot is shown in Fig. 2.3. We see that the measured results and the theoretical values fit nicely, even though the theory is for an infinite system size, and the simulations where performed at $L = 1000$. We also see that for larger values of s there are fewer observed values. It may therefore be a good idea to make the bins

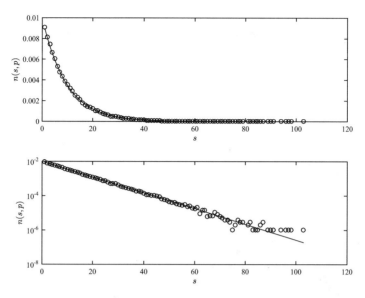

Fig. 2.3 Plot of the predicted $\overline{n(s, p)}$, based on $M = 1000$ samples of a $L = 1000$ system with $p = 0.9$, and the theoretical $n(s, p)$ curve on a linear scale (top) and a semilogarithmic scale (bottom). The semilogarithmic plot shows that $\overline{n(s, p)}$ follows an exponential curve

used for the histogram larger for larger values of s. We will return to this when we measure the cluster number density in two-dimensional systems in Chap. 4.

Average Cluster size

Since we have an exact expression for the cluster number density, $n(s, p)$ we can use it to calculate the average cluster size. However, what do we mean by the average cluster size in this case? In percolation theory it is common to define the average cluster size as the average size of a cluster connected to a given (random) site in our system. That is, we will use the cluster number density, $n(s, p)$, as the basic distribution for calculating the moments.

Average Cluster Size The average cluster size $S(p)$ is defined as

$$S(p) = \langle s \rangle = \sum_s s \left(\frac{sn(s, p)}{\sum_s sn(s, p)} \right), \qquad (2.30)$$

The normalization sum in the denominator is equal to p when $p < p_c$. In this case, we can therefore write this as

$$S(p) = \sum_s s\left(\frac{sn(s, p)}{p}\right) .$$

(2.31)

We can calculate the average cluster size from:

$$S = \frac{1}{p}\sum_s s^2 n(s, p) = \frac{(1 - p)^2}{p}\sum_s s^2 p^s$$

(2.32)

$$= \frac{(1 - p)^2}{p}\sum_s p\frac{\mathrm{d}}{\mathrm{d}p}p\frac{\mathrm{d}}{\mathrm{d}p}p^s = \frac{(1 - p)^2}{p}p\frac{\mathrm{d}}{\mathrm{d}p}p\frac{\mathrm{d}}{\mathrm{d}p}\sum_s p^s$$

(2.33)

$$= \frac{(1 - p)^2}{p}p\frac{\mathrm{d}}{\mathrm{d}p}p\frac{\mathrm{d}}{\mathrm{d}p}\frac{1}{1 - p} = (1 - p)^2\frac{\mathrm{d}}{\mathrm{d}p}\frac{p}{(1 - p)^2}$$

(2.34)

$$= (1 - p)^2\left(\frac{1}{(1 - p)^2} + \frac{2p}{(1 - p)^3}\right) = \frac{1 + p}{1 - p} ,$$

(2.35)

where we have used the trick introduced in (2.7) to move the derivation out through the sum.

This shows that we can write

$$S = \frac{1 + p}{1 - p} = \frac{\Gamma}{|p - p_c|^\gamma} ,$$

(2.36)

with $\gamma = 1$ and $\Gamma(p) = 1 + p$. That is, the average cluster size, S, also diverges as a power-law when p approaches p_c. The exponent $\gamma = 1$ of the power-law is again universal. That is, it depends on features such as dimensionality, but not on details such as the lattice structure.

2.3 Spanning Cluster

The density of the spanning cluster, $P(p; L)$, can be found using similar approaches. The spanning cluster only exists for $p \geq p_c$. The discussion for $P(p; L)$ is therefore not that interesting for the one-dimensional case. However, we can still introduce some of the general notions.

The behavior of $P(p; \infty)$ for $L \to \infty$ in one dimension is

$$P(p; \infty) = \begin{cases} 0 & \text{when } p < 1 \\ 1 & \text{when } p = 1 \end{cases} .$$

(2.37)

What is the relation between $P(p; L)$ and the distribution of cluster sizes? The distribution of the sizes of finite clusters is described by $sn(s, p)$, which is the probability that a given site belongs to a cluster of size s. If we look at a given site, that site is occupied with probability p. If a site is occupied it is either part of a finite cluster of size s or it is part of the spanning cluster. Since these two events cannot occur at the same time, the probability for a site to be occupied must be the sum of the probability to belong to a finite cluster and the probability to belong to the infinite cluster. The probability to belong to a finite cluster is the sum of the probabilities to belong to a cluster of s for all s. We therefore have the equality:

$$p = P(p; L) + \sum_s sn(s, p; L) , \qquad (2.38)$$

which is valid for percolation in any dimension, since we have not assumed anything about the dimensionality in this argument.

We can use this relation to find the density of the spanning cluster from the cluster number density $n(s, p)$ through

$$P(p) = p - \sum_s sn(s, p) . \qquad (2.39)$$

This illustrates that the cluster number density $n(s, p)$ is a fundamental property, which can be used to deduce many aspects of the percolation system.

2.4   Correlation Length

From the simulations in Fig. 1.4 we see that the size of the clusters increases as $p \to p_c$. We expect a similar behavior for the one-dimensional system. We have already seen that the mass (or area) of the clusters diverges as $p \to p_c$. The characteristic cluster size s_ξ characterizes the mass (or area) of a cluster. How can we characterize the *extent* of a cluster?

To characterize the linear extent of a cluster, we find the probability for two sites at a distance r to belong to the same cluster. This probability is called the **correlation function**, $g(r)$:

> **Correlation Function** The correlation function $g(r)$ describes the conditional probability that two sites a and b, which both are occupied and are separated by a distance r, belong to the same cluster.

For one-dimensional percolation, two sites a and b are part of the same cluster if and only if all the points in between a and b are occupied. If r denotes the number of points between a and b (not counting the start and end positions) as illustrated in

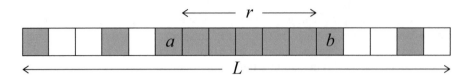

Fig. 2.4 An illustration of the distance r between two sites a and b. The two sites a and b are connected if and only if all the sites between a and b are occupied

Fig. 2.4, we find that the correlation function is

$$g(r) = p^r = e^{-r/\xi} , \qquad (2.40)$$

where $\xi = -\frac{1}{\ln p}$ is called the correlation length. The correlation length diverges as $p \to p_c = 1$. We can again find the way in which it diverges from a Taylor expansion in $(1 - p)$ when $p \to 1$

$$\ln p = \ln(1 - (1 - p)) \simeq -(1 - p) . \qquad (2.41)$$

We find that the correlation length is

$$\xi = \xi_0 (p_c - p)^{-\nu} , \qquad (2.42)$$

with $\nu = 1$. The correlation length therefore diverges as a power-law when $p \to p_c = 1$. This behavior is general for percolation theory, although the particular value of the exponent ν depends on the dimensionality.

We can use the correlation function to strengthen our interpretation of when a finite system size becomes relevant. As long as $\xi \ll L$, we will not notice the effect of a finite system, because no cluster is large enough to notice the finite system size. However, when $\xi \gg L$, the behavior is dominated by the system size L, and we are no longer able to determine how close we are to percolation. We will address these arguments in more detail in Chap. 5.

Exercises

Exercise 2.1 (Next-Nearest Neighbor Connectivity in 1d) Assume that connectivity is to the next-nearest neighbors for an infinite one-dimensional percolation system.

(a) Find $\Pi(p, L)$ for a system of length L.
(b) What is p_c for this system?
(c) Find $n(s, p)$ for an infinite system.

Exercise 2.2 (Higher Moments of s) The k'th moment of s is defined as

$$\langle s^k \rangle = \sum_s s^k \left(\frac{sn(s, p)}{p} \right) .$$ (2.43)

(a) Find the second moment of s as a function of p.

(b) Calculate the first moment of s numerically from $M = 1000$ samples for $p = 0.90, 0.95, 0.975$ and 0.99. Compare with the theoretical result.

(c) Calculate the second moment of s numerically from $M = 1000$ samples for $p = 0.90, 0.95, 0.975$ and 0.99. Compare with the theoretical result.

Infinite-Dimensional Percolation

3

In this chapter we address the percolation problem on an infinite-dimensional lattice without loops. In this case, it is possible to calculate several of the properties of the percolation system analytically. This allows us to develop a general theory and to develop concepts to be used for finite-dimensional systems. We introduce the infinite-dimensional Bethe lattice for a given coordination number. We find an exact solution for P and the average cluster size S, and use a Taylor-expansion to find an expression for $n(s, p)$. The methods and functional forms for $n(s, p)$ we introduce here, are used to interpret results in finite dimensions.

We have now seen how the percolation problem can be solved exactly for a one-dimensional system. However, in this case the percolation threshold is $p_c = 1$, and we were not able to address the behavior of the system for $p > p_c$. There is, however, another system in which many features of the percolation problem can be solved exactly. This is percolation on a regular tree structure on which there are no loops. The condition of no loops is essential. This is also why we call this system a system of infinite dimensions, because we need an infinite number of dimensions in Euclidean space in order to embed a tree without loops. In this section, we will provide an explicit solution to the percolation problem on a particular tree structure called the Bethe lattice [5].

The Bethe lattice, which is also called the Cayley tree, is a tree structure in which each node has Z neighbors. This structure has no loops. If we start from the central point and draw the lattice, the perimeter grows as fast as the bulk. Generally, we will call Z the coordination number. The Bethe lattice is illustrated in Fig. 3.1.

3.1 Percolation Threshold

If we start from the center in Fig. 3.1 and move along a branch, we find $(Z - 1)$ new neighbors from each of the branches. To get a spanning cluster, we need to ensure that at least one of the $Z - 1$ sites are occupied on average. That is, the occupation

© The Author(s) 2024
A. Malthe-Sørenssen, *Percolation Theory Using Python*, Lecture Notes in Physics 1029, https://doi.org/10.1007/978-3-031-59900-2_3

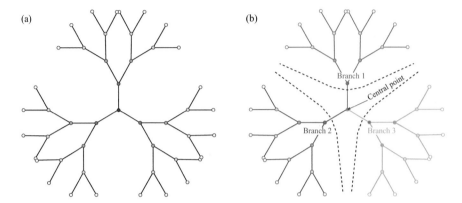

Fig. 3.1 Four generations of the Bethe lattice with coordination number $Z = 3$. (**a**) Illustration of a two-dimensional embedding of the lattice. (**b**) Illustration of how a central node is connected to three branches

probability, p, must be:

$$p(Z - 1) \geq 1 \, , \tag{3.1}$$

in order for this process to continue indefinitely.

We associate p_c with the value for p where the cluster is on the verge of dying out, that is

$$p_c = \frac{1}{Z - 1} \, . \tag{3.2}$$

For $Z = 2$ we regain the one-dimensional system, with percolation threshold $p_c = 1$. However, when $Z > 2$, we obtain a finite percolation threshold, that is, $p_c < 1$, which means that we can observe the behavior both above and below p_c.

In the following, we will demonstrate how we can find the density of the spanning cluster, $P(p)$, and the average cluster size S, before we address the scaling behavior of the cluster number density $n(s, p)$.

3.2   Spanning Cluster

We will use a standard trick to find the density $P(p)$ of the spanning cluster when $p > p_c$. The technique is based on starting from a "central" site, and then address the probability that a given branch is connected to infinity.

Relating P to $n(s, p)$ We start by noting that P is the probability for a site to be connected to the spanning cluster. If a site is present, which has a probability p, it must either belong to the infinite cluster with probability P or to one of the finite clusters with probability $\sum_s sn(s, p)$. This means that

$$p = P + \sum_s sn(s, p) , \qquad (3.3)$$

The sum $\sum_s sn(s, p)$ is the probability that the site is part of a finite cluster, which means that it is not connected to infinity. We introduce $Q(p)$ to denote the probability that a branch does not lead to infinity. The concept of a central point and a branch is illustrated in Fig. 3.1.

Deriving an Equation for $Q(p)$ If the probability that a site is not connected to infinity in a particular direction is Q, then the probability that the site is not connected to infinity in any direction is Q^Z. The probability that the site *is* connected to infinity is therefore $1 - Q^Z$. In addition, we need to include the probability p that the site is occupied. The probability that a given site is connected to infinity, that is, that it is part of the spanning cluster, is therefore

$$P = p(1 - Q^Z) . \qquad (3.4)$$

Now, we need to find an expression for $Q(p)$. We will determine Q through a consistency equation. Let us assume that we are moving along a branch, and that we have come to a point k. Then, Q gives the probability that this branch does not lead to infinity. This can occur either if site k is not occupied, which has a probability $(1 - p)$, or if site k is occupied, which has probability p, and all of the $Z - 1$ branches that lead out of k are not connected to infinity, which has probability Q^{Z-1}. The probability Q for the branch not to be connected to infinity is therefore

$$Q = (1 - p) + pQ^{Z-1} . \qquad (3.5)$$

We check this equation for the case $Z = 2$, which corresponds to a one-dimensional system. In this case we have $Q = 1 - p + pQ$, which gives, $(1 - p)Q = (1 - p)$, where we see that when $p \neq 1$, $Q = 1$. That is, when $p < 1$ all branches are not connected to infinity, implying that there is no spanning cluster. We therefore regain the results from one-dimensional percolation theory.

Solving to Find $Q(p)$ We could solve this equation for general Z. However, for simplicity we will restrict ourselves to $Z = 3$, which is the smallest Z that gives a behavior different from the one-dimensional system. We recall from (3.2) that for $Z = 3$, $p_c = 1/2$. We insert $Z = 3$ in (3.5), which gives

$$Q = 1 - p + pQ^2 , \qquad (3.6)$$

$$pQ^2 - Q + 1 - p = 0 . \qquad (3.7)$$

The solution of this second order equation is

$$Q = \frac{+1 \pm \sqrt{1 - 4p(1-p)}}{2p} = \frac{1 \pm \sqrt{(2p-1)^2}}{2p} = \frac{1 \pm |(2p-1)|}{2p} \ . \tag{3.8}$$

First, we notice that for $(2p-1) = 0$, that is for $p = 1/2$ (which is p_c for $Z = 3$), we find that $Q = 1/2p = 1$. Therefore, no branch propagates to infinity for $p = p_c = 1/2$. Second, for $(2p-1) < 0$, which corresponds to $p < 1/2 = p_c$, we get the solution

$$Q = \frac{1 \pm (1 - 2p)}{2p} \ , \tag{3.9}$$

which has two solutions: $Q = 1$ or $Q = (1-p)/p$. The second solution $(1-p)/p$ is greater than 1 when $p < 1/2$. We therefore conclude that for $p < 1/2$ and for $p = 1/2$, no branch propagates to infinity. This means that there is no infinite cluster (no spanning cluster) for $p \leq 1/2 = p_c$. Third, for $(2p-1) > 0$, which corresponds to $p > 1/2 = p_c$, we get the solution

$$Q = \frac{1 \pm (2p - 1)}{2p} \ , \tag{3.10}$$

which has two solutions: $Q = 1$ or $Q = (1-p)/p$. The second solution is smaller than 1 when $p > 1/2$. This means that there is a finite probability for a branch to propagate to infinity and for there to be a spanning cluster. We have therefore found that for $p \leq 1/2$, there is no spanning cluster, but for $p > 1/2$ there is a finite probability for a spanning cluster. This finding confirms that $1/2$ indeed is the percolation threshold for this system.

Finding $P(p)$ We insert $Q = (1-p)/p$ back into the equation for $P(p)$ to find the behavior of $P(p)$ for $p > p_c = 1/2$:

$$P = p(1 - Q^3) = p(1 - (\frac{1-p}{p})^3) \ . \tag{3.11}$$

This result is illustrated in Fig. 3.2. We notice that when $p \to 1$, we have that $(1-p)/p \to 0$. Therefore $P \propto p$ in this limit. To address the behavior when $p \to p_c = 1/2$, we use the identity $(1 - a^3) = (1 - a)(1 + a + a^2)$ and that $1 - (1-p)/p = 2(p - 1/2)$ to rewrite the expression in (3.11) to become

$$P = p(1 - (\frac{1-p}{p})^3) = p \left(1 - \frac{1-p}{p}\right) \left(1 + \frac{1-p}{p} + \left(\frac{1-p}{p}\right)^2\right) \ . \tag{3.12}$$

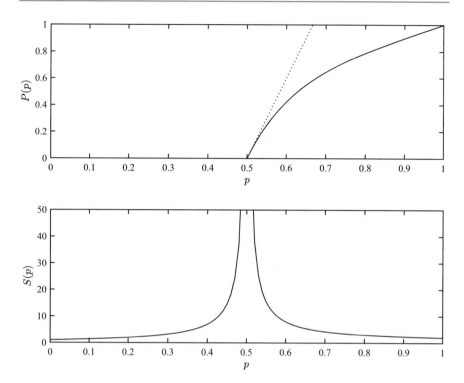

Fig. 3.2 (Top) A plot of $P(p)$ as a function of p for the Bethe lattice with $Z = 3$. The tangent at $p = p_c$ is illustrated by a straight line. (Bottom) A plot of the average cluster size, $S(p)$, as a function of p for the Bethe lattice with $Z = 3$. The average cluster size diverges when $p \to p_c = 1/2$ both from below and above

We can rewrite this as

$$P = 2\left(p - \frac{1}{2}\right)\left(1 + \frac{1-p}{p} + \left(\frac{1-p}{p}\right)^2\right) . \tag{3.13}$$

We Taylor expand the second term around $p = p_c = 1/2$:

$$\left(1 + \frac{1-p}{p} + \left(\frac{1-p}{p}\right)^2\right) \simeq 3 + \mathcal{O}(p - p_c) , \tag{3.14}$$

which when inserted back into (3.13) gives

$$P \simeq 6(p - p_c) + \mathcal{O}\left((p - p_c)^2\right) . \tag{3.15}$$

Since we are only interested in the leading term, we have found that for $p > p_c = 1/2$ we can approximate $P(p)$ as:

$$P(p) \simeq B(p - p_c)^\beta , \tag{3.16}$$

where $B = 6$ and the exponent $\beta = 1$. The density of the spanning cluster, P, is therefore a power-law in $(p - p_c)$ with exponent β. In general, this exponent depends on the dimensionality of the lattice, but it does not depend on lattice details, such as the number of neighbors Z. We will leave it as an exercise for the reader to show that β is the same for $Z = 4$.

The approach we have used here is often called a mean field solution or a self-consistency solution: We assume that we know Q, and then solve to find Q. We will use similar methods later.

3.3   Average Cluster Size

We will use a similar method to find the average cluster size, $S(p)$. We start by defining $T(p)$ as the average number of sites connected to a given site on a specific branch, such as in branch 1 in Fig. 3.1. The average cluster size S is then given as

$$S = 1 + ZT , \tag{3.17}$$

where the 1 represents the central point, T is the average number of sites on each branch and Z is the number of branches. We will again attempt to find a self-consistent solution for T, starting from a center site. The average cluster size T is found from summing the probability that the next site k is empty, $(1 - p)$, multiplied with the contribution to the average, in this case (0), plus the probability that the next site is occupied, p, multiplied with the contribution in this case, which is the contribution from the site (1) and the contribution of the remaining $Z - 1$ subbranches. In total:

$$T = (1 - p)\,0 + p\,(1 + (Z - 1)T) , \tag{3.18}$$

We solve for T, finding

$$T = \frac{p}{1 - p(Z - 1)} . \tag{3.19}$$

This expression diverges when $1 - p(Z - 1) = 0$, that is, for $p = 1/(Z - 1)$, which we recognize as $p_c = 1/(Z - 1)$. We insert this in (3.17) as find that S is

$$S = 1 + ZT = \frac{1 + p}{1 - (Z - 1)p} = \frac{1 + p}{1 - \frac{p}{p_c}} = \frac{p_c(1 + p)}{p_c - p} , \tag{3.20}$$

which is illustrated in Fig. 3.2. This is a special case for $S(p)$, which in general can be written on the general form

$$S = \frac{\Gamma}{(p_c - p)^\gamma} \, . \tag{3.21}$$

For the Bethe lattice with coordination number Z, we have found that $p_c = 1/(Z - 1)$ and that the exponent is $\gamma = 1$. where our argument determines $p_c = 1/(Z - 1)$, and the exponent $\gamma = 1$. The average cluster size S therefore diverges as a power-law when p approaches p_c. The exponent γ characterizes the behavior. This is a general results. The value of γ depends on the dimensionality, but not on the details of the lattice. Here, we notice that γ does indeed not depend on Z.

3.4   Cluster Number Density

In order to find the cluster number density for the Bethe lattice, we start by developing a more general way to find the cluster number density. To find the cluster number density for a given s, we need to find all possible configurations, $c(s)$, of clusters of size s, and sum up their probability:

$$n(s, p) = \sum_{c(s)} p^s (1 - p)^{t(c)} \, . \tag{3.22}$$

This was simple in one dimension, because there was only one possible configuration for a given s. Just like in one dimension we include the term p^s, because we know that we must have all the s sites of the cluster present. In addition, we need to include a term that takes into account that all the neighboring sites must be empty. In general, we introduce the notation $t(c)$ for the number of neighbors for configuration c. In one dimension, t is always 2, but in higher dimensions different clusters may have different numbers of neighbors, as illustrated for the case of a two-dimensional system in Fig. 3.3. We therefore need to include the term $(1 - p)^t$ to account for the probability for the t neighbors to be unoccupied. Based on this, we realize that we may sum over all values of t. However, we then need to include the effect that there are several clusters that can have the same t. We will then have

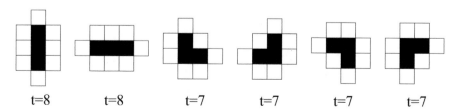

Fig. 3.3 Illustration of the 6 possible configurations for a two-dimensional cluster of size $s = 3$

to introduce a degeneracy factor $g_{s,t}$ which gives the number of different clusters that have size s and a number of neighbors equal to t. The cluster number density can then be written as

$$n(s, p) = p^s \sum_t g_{s,t}(1 - p)^t . \tag{3.23}$$

Degeneracy for Two-Dimensional Clusters We can illustrate these concept for two-dimensional percolation. Let us study the case when $s = 3$. In this case there are 6 possible clusters for size $s = 3$, as illustrated in Fig. 3.3. There are two clusters with $t = 8$, and four clusters with $t = 7$. There are no other clusters of size $s = 3$. We can therefore conclude that for the two-dimensional lattice, we have $g_{3,8} = 2$, and $g_{3,7} = 4$, and $g_{3,t} = 0$ for all other values of t.

Degeneracy for the Bethe Lattice For the Bethe lattice, there is a particularly simple relation between the number of sites, s, and the number of neighbors, t. We can see this by looking at the first few generations of a Bethe lattice grown from a central seed. For $s = 1$, the number of neighbors is $t_1 = Z$. To add one more site, we have to remove one neighbor from what we had previously, and then we add $Z - 1$ new neighbors, that is, for $s = 2$ we have $t_2 = t_1 + (Z - 2)$. Consequently we get an iterative relation for the number of neighbors t_k when we have k sites:

$$t_k = t_{k-1} + (Z - 2) , \tag{3.24}$$

and therefore:

$$t_s = s(Z - 2) + 2 . \tag{3.25}$$

Cluster Number Density The cluster number density, given by the sum over all t, is therefore reduced to only a single term for the Bethe lattice

$$n(s, p) = g_{s,t_s} p^s (1 - p)^{t_s} , \tag{3.26}$$

For simplicity, we will write $g_s = g_{s,t_s}$. In general, we do not know g_s, but we will show that we still can learn quite a lot about the behavior of $n(s, p)$. The cluster density can therefore be written as

$$n(s, p) = g_s p^s (1 - p)^{2+(Z-2)s} . \tag{3.27}$$

We rewrite this as a common factor to the power s:

$$n(s, p) = g_s [p(1 - p)^{Z-2}]^s (1 - p)^2 , \tag{3.28}$$

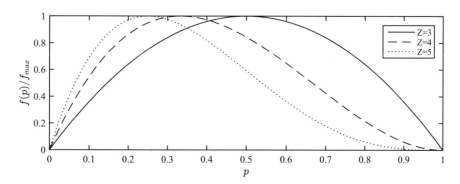

Fig. 3.4 A plot $f(p) = p(1-p)^{Z-2}$, which is a term in the cluster number density $n(s, p) = g_s[p(1-p)^{Z-2}]^s(1-p)^2$ for the Bethe lattice. We notice that $f(p)$ has a maximum at $p = p_c$, and that the second derivative, $f''(p)$, is zero in this point. A Taylor expansion of $f(p)$ around $p = p_c$ will therefore have a second order term in $(p-p_c)$ as the lowest-order term—to lowest order it is a parabola at $p = p_c$. It is this second order term which determines the exponent σ, which consequently is independent of Z

which in the special case for $Z = 3$, which we studied above, becomes

$$n(s, p) = g_s[p(1-p)]^s(1-p)^2 \, . \tag{3.29}$$

Taylor Expansion Around $p = p_c$ Let us now address $n(s, p)$ for p close to p_c for a general value of Z. In this range, we will do a Taylor expansion of the term $f(p) = p(1-p)^{Z-2}$, which is raised to the power s in the equation for $n(s, p)$ in (3.28). The shape of $f(p)$ as a function of p is shown in Fig. 3.4. The maximum of $f(p)$ occurs for $p = p_c = 1/(Z-1)$. This is also easily seen from the first derivative of $f(p)$.

$$f'(p) = (1-p)^{Z-2} - p(Z-2)(1-p)^{Z-3} \tag{3.30}$$

$$= (1-p)^{Z-3}(1-p-p(Z-2)) \tag{3.31}$$

$$= (1-p)^{Z-3}(1-(Z-1)p) \tag{3.32}$$

which shows that $f'(p_c) = 0$. (We leave it to the reader to show that $f''(p_c) < 0$.) The Taylor expansion of $f(p)$ around $p = p_c$ is then:

$$f(p) = f(p_c) + f'(p_c)(p-p_c) + \frac{1}{2}f''(p_c)(p-p_c)^2 + \mathcal{O}((p-p_c)^3) \, , \tag{3.33}$$

where we already have found the first order term, $f'(p_c) = 0$. We can therefore write

$$f(p) \simeq f(p_c) - \frac{1}{2}f''(p_c)(p-p_c)^2 = A(1 - B(p-p_c)^2) \, . \tag{3.34}$$

Cluster Number Density We will now insert this Taylor expansion back into the expression for the cluster number density:

$$n(s, p) = g_s[f(p)]^s (1 - p)^2 = g_s e^{s \ln f(p)} (1 - p)^2 , \qquad (3.35)$$

where we now insert $f(p) \simeq A(1 - B(p - p_c)^2)$ to get

$$n(s, p) \simeq g_s A^s e^{s \ln(1 - B(p - p_c)^2)} (1 - p)^2 . \qquad (3.36)$$

When p is close to p_c, $(p - p_c)^2$ is small number, and we can use the first order of the Taylor expansion of $\ln(1 - x) \simeq -x + \mathcal{O}(x^2)$, to get

$$n(s, p) \simeq g_s A^s e^{-sB(p - p_c)^2} (1 - p)^2 . \qquad (3.37)$$

Consequently, for $p = p_c$ we get

$$n(s, p_c) = g_s A^s (1 - p_c)^2 . \qquad (3.38)$$

Cluster Number Density Expressed in Terms of $n(s, p_c)$ When p is close to p_c, we can assume that $(1 - p)^2 \simeq (1 - p_c)^2$, and we can therefore rewrite the cluster number density in terms of $n(s, p_c)$, giving

$$n(s, p) = n(s, p_c) e^{-sB(p - p_c)^2} . \qquad (3.39)$$

We rewrite the exponential in terms of a characteristic cluster size s_ξ as

$$n(s, p) = n(s, p_c) e^{-s/s_\xi} , \qquad (3.40)$$

where the characteristic cluster size s_ξ is

$$s_\xi = B^{-1}(p - p_c)^{-2} . \qquad (3.41)$$

This implies that the characteristic cluster size s_ξ diverges as a power-law with exponent $1/\sigma = 2$ as p approaches p_c. The general scaling form for the characteristic cluster size s_ξ is

$$s_\xi \propto |p - p_c|^{-1/\sigma} , \qquad (3.42)$$

where the exponent σ is universal, which means that is does not depend on lattice details such a Z, but it does depend on lattice dimensionality. It will therefore have a different value for two-dimensional percolation.

Scaling Ansatz for $n(s, p_c)$ We can use our knowledge of the behavior of P and S when p approaches p_c to constrain the scaling behavior of $n(s, p_c)$. We know that we can find S and P from the cluster number density. The average cluster size S as p approaches p_c is

$$S = \frac{\Gamma}{p_c - p} ,$$ (3.43)

which should diverge when p approaches p_c. We rewrite S in terms of $n(s, p)$:

$$S = \sum s^2 n(s, p) ,$$ (3.44)

which should diverge as p approaches p_c. We rewrite the sum as an integral in the limit of $p = p_c$

$$S = \int_0^\infty s^2 n(s, p_c) ds .$$ (3.45)

This integral must diverge. We can therefore conclude that $n(s, p_c)$ is not an exponential, otherwise the integral then would converge. We therefore assume that the cluster number density has a power-law shape, that is, we introduce the scaling ansatz:

$$n(s, p_c) \simeq C s^{-\tau} .$$ (3.46)

This expression is only valid in the limit when $s \gg 1$. We introduce this scaling ansatz into the expression for P:

$$\sum_s s n(s, p) = p - P ,$$ (3.47)

and approximate the sum with an integral in the limit when $p = p_c$:

$$\int_0^\infty s n(s, p_c) ds = \int_0^\infty s C s^{-\tau} ds = p - P ,$$ (3.48)

which should converge. We therefore have two conditions

1. The integral

$$\int_0^\infty s^2 n(s, p_c) ds = \int_0^\infty s^2 C s^{-\tau} ds$$ (3.49)

should diverge, which implies that $\tau - 2 \leq 1$

1. The integral

$$\int_0^\infty sn(s, p_c)ds = \int_0^\infty sCs^{-\tau}ds \tag{3.50}$$

should converge, which implies that $\tau - 1 > 1$.

This implies that the exponent τ has the following bounds:

$$2 < \tau \le 3 . \tag{3.51}$$

Scaling Behavior of $S(p)$ When p Is Close to p_c We can sum up our arguments so far in the relation

$$n(s, p) = n(s, p_c)e^{-B(p-p_c)^2 s} = Cs^{-\tau}e^{-B(p-p_c)^2 s} = Cs^{-\tau}e^{-s/s_\xi} . \tag{3.52}$$

Let us use this expression to calculate S, for which we know the exact scaling behavior, and then again use this to find the value for τ

$$S = C\sum_s s^{2-\tau}e^{-s/s_\xi} \rightarrow C\int_1^\infty s^{2-\tau}e^{-s/s_\xi}ds . \tag{3.53}$$

We now make a rough estimate. This is useful, since it is in the spirit of this book, and it also provides the correct behavior. We assume that

$$S = C\int_1^\infty s^{2-\tau}e^{-s/s_\xi}ds \sim C\int_1^{s_\xi} s^{2-\tau}ds \sim s_\xi^{3-\tau} , \tag{3.54}$$

where we have used the sign $\sim$ to denote that the expressions have the same scaling behavior. We can do it slightly more elaborately:

$$S \simeq C\int_1^\infty s^{2-\tau}e^{-s/s_\xi}ds . \tag{3.55}$$

We change variables by introducing, $u = s/s_\xi$, which gives

$$S \simeq s_\xi^{3-\tau}\int_{1/s_\xi}^\infty u^{2-\tau}e^{-u}du . \tag{3.56}$$

This integral is simply a number, since $1/s_\xi \rightarrow 0$, when $p \rightarrow p_c$. The asymptotic scaling behavior in the limit $p \rightarrow p_c$ is therefore

$$S \sim s_\xi^{3-\tau} \sim (p - p_c)^{-2(3-\tau)} \sim (p - p_c)^{-1} , \tag{3.57}$$

where we have used that

$$s_\xi \sim (p - p_c)^{-2} \, , \tag{3.58}$$

and that

$$S \sim (p - p_c)^{-1} \, . \tag{3.59}$$

Our direct solution therefore gives that

$$\tau = \frac{5}{2} \, . \tag{3.60}$$

This relation indeed satisfies the exponent relations we found above, since $2 <$ $5/2 \leq 3$. A plot of the scaling form is shown in Fig. 3.5.

Preliminary Scaling Theory for Cluster Number Density   We have now developed a preliminary scaling theory for the cluster number density. In the coming chapters, we will demonstrate that similar scaling relations also are valid for percolation in other dimension. We have found that in the vicinity of p_c, we do not expect deviations until we reach large s, that is, until we reach a characteristic cluster size s_ξ that increases as $p \to p_c$. The general scaling form for the cluster number density is

$$n(s, p) = n(s, p_c)F(\frac{s}{s_\xi}) \, , \tag{3.61}$$

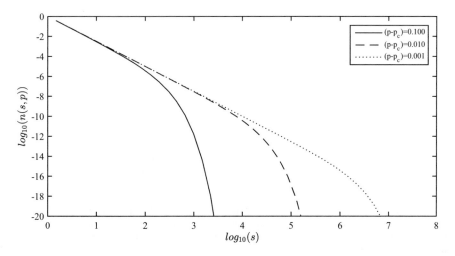

Fig. 3.5 A plot of $n(s, p) = s^{-\tau} \exp(-s(p - p_c)^2)$ as a function of s for various values of p illustrates how the characteristic cluster size s_ξ appears as a cut-off in the cluster number density

where

$$n(s, p_c) = Cs^{-\tau} , \tag{3.62}$$

and

$$s_\xi = s_0 |p - p_c|^{-1/\sigma} . \tag{3.63}$$

In addition, we have the following scaling relations:

$$P(p) \sim (p - p_c)^\beta , \tag{3.64}$$

$$\xi \sim |p - p_c|^{-\nu} , \tag{3.65}$$

and

$$S \sim |p - p_c|^{-\gamma} , \tag{3.66}$$

with a possible non-trivial behavior for higher moments of the cluster density.

Exercises

Exercise 3.1 ($P(p)$ for $Z = 4$) Find $P(p)$ for $Z = 4$ and determine β for this value of Z.

Finite-Dimensional Percolation

4

In this chapter we apply the scaling theories developed in the one-dimensional system and the infinite-dimensional system to systems of finite dimensions. The lowest dimension with an interesting behavior is two dimensions. Here, we introduce effective ways to measure the cluster number density $n(s, p)$ in two dimension. We develop the scaling theory for $n(s, p)$ and demonstrate how to use data-collapse plots as an efficient method to measure the critical exponents. We also demonstrate how we can use the scaling theory for $n(s, p)$ to derive expressions for the density of the spanning cluster, P, and the average cluster size, S. Finally, we demonstrate how the scaling theory provides scaling relations, that is, relations between exponents, and bounds for the values of the critical exponents.

For the one-dimensional and the infinite-dimensional systems we have been able to find exact results for the percolation probability, $\Pi(p)$, for $P(p)$, the probability for a site to belong to an infinite cluster, and we have characterized the behavior using the distribution of cluster sizes, $n(s, p)$ and its cut-off, s_ξ. In both one and infinite dimensions we have been able to calculate these functions exactly. However, in two and three dimensions—which are the most relevant for our world—we are unfortunately not able to find exact solutions. We saw above that the number of configurations in a L^d system in d-dimensions increases very rapidly with L—so rapidly that a complete enumeration is impossible. But can we still use what we learned from the one and infinite-dimensional systems?

In the one-dimensional case it was simple to find $\Pi(p, L)$ because there is only one possible path from one side to another. We cannot generalize this to two dimensions, since in two dimensions there are many paths from one side to another—and we need to include all to estimate the probability for percolation. Similarly, it was simple to find $n(s, p)$ in one dimension, because all clusters only have two neighboring sites and the surface, t, is always of size 2. This is also not generalizable to higher dimensions.

In the infinite-dimensional system, that is in the Bethe lattice, we were able to find $P(p)$ because we could separate the cluster into different paths that never could

© The Author(s) 2024   47
A. Malthe-Sørenssen, *Percolation Theory Using Python*, Lecture Notes
in Physics 1029, https://doi.org/10.1007/978-3-031-59900-2_4

intersect except in a single point, because there are no loops in the Bethe lattice. This is not the case in two and three dimensions, where loops always will be possible. When there are loops present, we cannot use the arguments we used for the Bethe lattice, because a branch cut off at one point may be connected again further out. For the Bethe lattice, we could also estimate the multiplicity $g(s, t)$ of the clusters, the number of possible clusters of size s and surface t, since t was a function of s. In a two- or three-dimensional system this is not similarly simple, because the multiplicity $g(s, t)$, that is the number of different cluster configurations with size s and surface t, is not simple even in two dimensions, as illustrated in Fig. 4.1.

This means that the solution methods used for the one dimensional and the infinite dimensional systems cannot be extended to address two-dimensional or three-dimensional systems. However, several of the techniques and observations we have made for the one-dimensional and the Bethe lattice systems, can be used as the basis for a generalized theory that can be applied in any dimension. Here, we will therefore pursue the more general features of the percolation system, starting with the cluster number density, $n(s, p)$.

4.1 Cluster Number Density

We have found that the cluster number density plays a fundamental role in our understanding of the percolation problem, and we will use it here as our basis for the scaling theory for percolation.

When we discussed the Bethe lattice, we found that we could write the cluster number density as a sum over all possible configurations of cluster size, s:

$$n(s, p) = \sum_j p^s (1 - p)^{t_j} , \tag{4.1}$$

where j runs over all different configurations, and t_j denotes the number of neighbors for this particular configuration. We can simplify this by rewriting the sum to be over all possible number of neighbors, t, and include the degeneracy $g_{s,t}$, the number of configurations with t neighbors:

$$n(s, p) = \sum_t g_{s,t} p^s (1 - p)^t . \tag{4.2}$$

The values of $g_{s,t}$ can be found for smaller values of s. However, while this may give us interesting information about the smaller cluster, and therefore for smaller values of p, it does not help us to develop a theory for the behavior for p close to p_c.

In order to address the cluster number density, we will need to study the characteristics of $n(s, p)$, for example by generating numerical estimates for its scaling behavior, and then propose a general scaling form which will be tested in various settings.

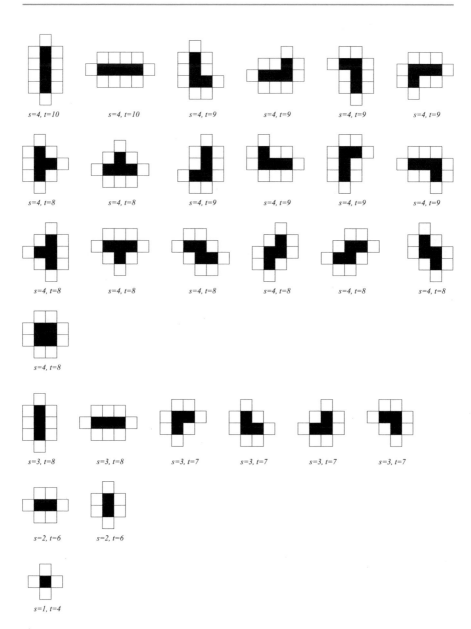

Fig. 4.1 Illustration of the possible configurations for two-dimensional clusters of size $s = 1, 2, 3, 4$

Numerical Estimation of $n(s, p)$

We discussed how to measure $n(s, p)$ from a set of numerical simulations in Chap. 2.
We can use the same method in two and higher dimensions. We estimate $n(s, p; L)$
using

$$\overline{n(s, p; L)} = \frac{N_s}{M \cdot L^d} \,, \tag{4.3}$$

where N_s is the total number of clusters of size s measured for M simulations in a
system of size L^d and for a given value of p. We perform these simulations just as
we did in one dimension, using the following program:

```
import numpy as np
import matplotlib.pyplot as plt
from scipy.ndimage import measurements
M = 2000
L = 200
p = 0.58
allarea = np.array([])
for i in range(M):
    z = np.random.rand(L,L)
    m = z<p
    lw, num = measurements.label(m)
    labelList = np.arange(lw.max() + 1)
    area = measurements.sum(m, lw, labelList)
    allarea = np.append(allarea,area)
n,sbins = np.histogram(allarea,bins=int(max(allarea)))
s = 0.5*(sbins[1:]+sbins[:-1])
nsp = n/(L*nsamp)
i = np.nonzero(n)
plt.figure(figsize=(12,4))
plt.subplot(1,2,1)
plt.plot(s[i],nsp[i],'o')
plt.xlabel('$s$')
plt.ylabel('$n(s,p)$')
plt.subplot(1,2,2)
plt.loglog(s[i],nsp[i],'o')
plt.xlabel('$s$')
plt.ylabel('$n(s,p)$')
```

The resulting plot of $\overline{n(s, p; L)}$ for $L = 200$ is shown in Fig. 4.2. Unfortunately,
this plot is not very useful. The problem is that there are too many values of s for
which we have little or no data at all. For small values of s we have many clusters
for each value of s and the statistics is good. But for large values of s, such as for
clusters of size $s = 10^4$ and above, we have less than one data point for each value
of s. Our measured distribution $\overline{n(s, p; L)}$ is therefore a poor representation of the
real $n(s, p; L)$ in this range.

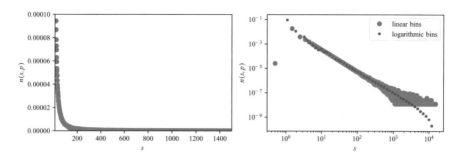

Fig. 4.2 Plot of $n(s, p; L)$ estimated from $M = 1000$ samples for $p = 0.58$ and $L = 200$. (Left) Direct plot. (Right) Log-log plot with linear and logaritmic binning

Measuring Probability Densities of Rare Events

The problem with the measured results in Fig. 4.2 occur because we have chosen a very small bin size for the histogram. For small values of s we want to have a small bin size, since the statistics here is good, but for large values of s we want to have larger bin sizes. This is often solved by using logarithmic binning: We make the bin edges a^i, where a is the basis for the bins and i is bin number. If we chose $a = 2$ as the basis for the bins, the bin edges will be $2^0, 2^1, 2^2, 2^3, \ldots$, that is 1, 2, 4, 8, $\ldots$. We then count how many events occur in each such bin. If we number the bins using the index i, then the edges of the bins are $s_i = a^i$, and the width of bin i is $\Delta s_i = s_{i+1} - s_i$. We then count how many events, N_i, occur in the range from s_i to $s_i + \Delta s_i$. The average number of clusters, $\bar{N}_i$ in each bin in the interval Δs_i is $\bar{N}_i = N_i/\Delta s_i$ for a single realization and $\bar{N}_i = N_i/(\Delta s_i M)$ for M realizations. The estimate for the cluster number density in the middle point of the bin, that is for $\bar{s}_i = (s_{i+1} + s_i)/2$, is

$$\overline{n(\bar{s}_i, p; L)} = \frac{\bar{N}_i}{L^d} = \frac{N_i}{M \Delta s_i L^d} \ . \tag{4.4}$$

A common mistake is to forget to divide by the bin size Δs_i when the bin sizes are not all the same! We implement method by generating an array of all the bin edges. First, we find an upper limit to the bins, that is, we find an i_m so that

$$a^{i_m} > \max(s) \ \Rightarrow \ \log_a a^{i_m} > \log_a \max(s) \ , \tag{4.5}$$

$$i_m > \log_a \max(s) \ . \tag{4.6}$$

We can for example round the right hand side up to the nearest integer

```
a = 1.2
logamax = np.ceil(np.log(max(allarea))/np.log(a));
```

where `allarea` corresponds to all the s-values. We can then generate an array of indices from 1 to this maximum value

```
logbins = a**np.arange(0,logamax)
```

And we can further generate the histogram with this set of bin edges

```
nl,nlbins = np.histogram(allarea,bins=logbins)
```

And calculate the bin sizes and the bin centers

```
ds = np.diff(logbins)
sl = 0.5*(logbins[1:]+logbins[:,-1])
```

Finally, we calculate the estimated value for $\overline{n(s, p; L)}$:

```
nsl = nl/(M*L**2*ds)
```

The complete code for this analysis is found in the following script

```
a = 1.2
logamax = np.ceil(np.log(max(s))/np.log(a))
logbins = a**np.arange(0,logamax)
nl,nlbins = np.histogram(allarea,bins=logbins)
ds = np.diff(logbins)
sl = 0.5*(logbins[1:]+logbins[:-1])
nsl = nl/(M*L**2*ds)
plt.loglog(sl,nsl,'.b')
```

The resulting plot for $a = 1.2$ is shown in Fig. 4.2. Notice that the logarithmically binned plot is much easier to interpret than the linearly binned plot. You should, however, always reflect on whether your binning method may influence the resulting plot in some way, since there may be cases where your choice of binning method may affect the results you get. Although this is not expected to play any role in your measurements in this book. We will in the following adapt logarithmic binning strategies whenever we measure a dataset which is sparse.

Measurements of $n(s, p)$ When $p \to p_c$

What happens to $n(s, p; L)$ when p is close to p_c? We perform a sequence of simulations for various values of p_c and plot the resulting values for $\overline{n(s, p; L)}$. The resulting plot is shown in Fig. 4.3.

Since the plot is double-logarithmic, a straight line corresponds to a power-law behavior, $n(s, p) \propto s^{-\tau}$. We see that as p approaches p_c the cluster number density $n(s, p)$ approaches a power-law. We see that the $n(s, p)$ curve follows the power-law behavior over some range of s-values, but drops rapidly for larger s-values. This is an effect of the characteristic cluster size, which also can be visually observed in Figs. 1.4 and 1.5, where we see that the characteristic cluster size increases as p approaches p_c. How can we characterize the characteristic cluster size based on this measurement of $n(s, p)$? We could measure s_ξ directly from the plot, by drawing

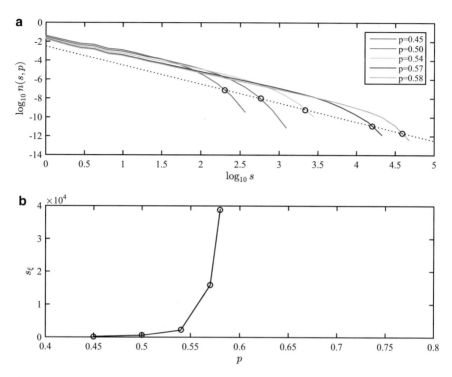

Fig. 4.3 (**a**) Plot of $n(s, p; L)$ as a function of s for various values of p for a 512×512 lattice. (**b**) Plot of $s_\xi(p)$ measured from the plot of $n(s, p)$ corresponding to the points shown in circles in (**a**)

a straight line parallel to but below $n(s, p_c)$, as illustrated in Fig. 4.3. When the measured, $\overline{n(s, p)}$ intersects this drawn line, $n(s, p)$ has fallen by a constant factor below $n(s, p_c)$. We *define* this as s_ξ, and we measure it by reading the values from the s-axis. The resulting set of s_ξ values are plotted as a function of p in Fig. 4.3. We see that s_ξ increases and possibly diverges as p approaches p_c. This is an effect we also found in the one-dimensional and the infinite-dimensional case, where we found that

$$s_\xi \propto |p - p_c|^{-1/\sigma} \tag{4.7}$$

where σ was 1 is one dimension. We will now use this to develop a theory for both $n(s, p; L)$ and s_ξ based on our experience from one and infinite dimensional percolation.

Scaling Theory for $n(s, p)$

When we develop a theory, we realize that we are only interested in the limit $p \to p_c$, that is $|p - p_c| \ll 1$, and $s \gg 1$. In this limit, we expect s_ξ to mark the crossover between two different behaviors. There is a common behavior for $n(s, p)$ for all p-values for small s, up to a cut-off, s_ξ, as we also observe in Fig. 4.3: The curves for different p-values are approximately equal for small s.

Based on what we observed in one-dimension and infinite-dimensions, we expect and propose the following scaling form for $n(s, p)$:

$$n(s, p) = n(s, p_c)F(\frac{s}{s_\xi}) , \qquad (4.8)$$

$$n(s, p_c) = Cs^{-\tau} , \qquad (4.9)$$

$$s_\xi = s_0|p - p_c|^{-1/\sigma} . \qquad (4.10)$$

Based on the methods presented in this book, we have estimated the exponents for various systems and listed them in Table 4.1. You can find an up-to-date list of all the exponents in the wikipedia article on percolation thresholds at https://en.wikipedia.org/wiki/Percolation_critical_exponents.

We will often simplify the scaling form by writing it on the form:

$$n(s, p) = s^{-\tau} F(s/s_\xi) = s^{-\tau} F((p - p_c)^{1/\sigma}s) . \qquad (4.11)$$

What can we expect from the scaling function $F(x)$?

This is essentially the prediction of a data-collapse. If we plot $s^\tau n(s, p)$ as a function of $s|p - p_c|^{1/\sigma}$ we would expect to get the scaling function $F(x)$, which should be a universal curve, as illustrated in Fig. 4.4.

An alternative scaling form is

$$n(s, p) = s^{-\tau} \hat{F}((p - p_c)s^\sigma) , \qquad (4.12)$$

where we have introduced the function $\hat{F}(u) = F(u^\sigma)$. These forms are equivalent, but in some cases this form produces simpler calculations.

This scaling form should in particular be valid for both the 1d and the Bethe lattice cases—let us check this in detail.

Table 4.1 Values for scaling exponents for percolation in 1, 2, 3, 4 and infinite dimensions [8, 37]

d	β	τ	σ	γ	ν	D	μ	D_{min}	D_{max}	D_B
1		2	1	1	1					
2	0.14	2.05	0.4	2.4	1.33	1.89	1.3	1.1	1.4	1.6
3	0.4	2.2	0.45	1.8	0.9	2.5	2.0	1.3	1.6	1.7
Bethe	1	5/2	1/2	1	1/2	4	3	2	2	2

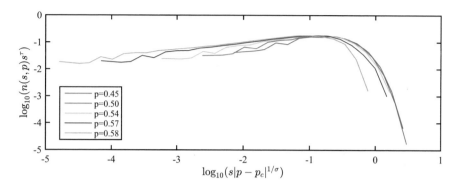

Fig. 4.4 A plot of $n(s, p)s^\tau$ as a function of $|p - p_c|^{1/\sigma}s$ shows that the cluster number density satisfies the scaling ansatz of (4.11)

Scaling Ansatz for 1d Percolation

In the case of one-dimensional percolation, we know that we can write the cluster density exactly as

$$n(s, p) = (1 - p)^2 e^{-s/s_\xi} . \tag{4.13}$$

We showed that we could rewrite this as

$$n(s, p) = s^{-2} F\left(\frac{s}{s_\xi}\right) , \tag{4.14}$$

where $F(u) = u^2 e^{-u}$. This is indeed on the general scaling form with $\tau = 2$.

Scaling Ansatz for Bethe Lattice

For the Bethe lattice we found that the cluster density was approximately on the form

$$n(s, p) \propto s^{-\tau} e^{-s/s_\xi} , \tag{4.15}$$

which is already on the wanted form, so that

$$n(s, p) = s^{-\tau} F(s/s_\xi) . \tag{4.16}$$

4.2   Consequences of the Scaling Ansatz

While the scaling ansatz has a simple form, it has powerful consequences. Here, we address the theoretical consequences of the scaling ansatz, and demonstrate how we can use the scaling ansatz in theoretical arguments. The methods we introduce here are important methods in scaling theories, and we will use them in theoretical arguments throughout this text.

Average Cluster Size

Let us demonstrate how we can use the scaling ansatz to calculate the scaling of the average cluster size, S, and how this can be used to provide limits for the exponent τ.

Definition of Average Cluster size S   The average cluster size, S, is defined as follows: We point to a random point in the percolation system. What is the average size of the cluster connected to that point? The probability that a random point is part of the cluster of size s is $sn(s, p)$ and the size of that cluster is s. We find the average cluster by summing over all (finite) clusters, that is from $s = 1$ to infinity:

$$S(p) = \sum_{s=1}^{\infty} ssn(s, p) = \sum_{s=1}^{\infty} s^2 n(s, p) \ . \tag{4.17}$$

We assume that we study systems where p is close to p_c so that the cluster number density $n(s, p)$ is wide and that its drop-off (crossover) s_ξ is rapid. The sum over s will then be a sum with many non-negligible terms and we can approximate this sum by an integral over s instead:

$$S(p) = \sum_{s=1}^{\infty} s^2 n(s, p) \simeq \int_1^{\infty} s^2 n(s, p) \, ds \ . \tag{4.18}$$

We can now insert the scaling ansatz $n(s, p) = s^{-\tau} F(s/s_\xi)$, getting:

$$S(p) = \int_1^{\infty} s^{2-\tau} F(s/s_\xi) \, ds \ , \tag{4.19}$$

We know that the function $F(s/s_\xi)$ goes very rapidly to zero when s is larger than s_ξ, and that it is approximately a constant when s is smaller than s_ξ. We will therefore approximate $F(u)$ by a step function which is a constant up to $u = 1$ and then 0 for $u > 1$. Consequently, we only integrate up to s_ξ, over a region where $F(s/s_\xi)$ is approximately a constant:

$$S(p) = \int_1^\infty s^{2-\tau} F(s/s_\xi) \, ds \simeq \int_1^{s_\xi} C s^{2-\tau} \, ds \ . \qquad (4.20)$$

We solve this integral, finding that

$$S(p) = C' s_\xi^{3-\tau} \ , \qquad (4.21)$$

where C' is a constant. We insert $s_\xi = |p - p_c|^{-1/\sigma}$, giving:

$$S(p) \propto \left(|p - p_c|^{-1/\sigma} \right)^{3-\tau} \propto |p - p_c|^{\frac{3-\tau}{\sigma}} \ . \qquad (4.22)$$

We recall that γ we is the scaling exponent of $S(p)$: $S(p) \propto |p - p_c|^{-\gamma}$. We have therefore found what we call a *scaling relation* between exponents:

$$\gamma = \frac{3 - \tau}{\sigma} \ . \qquad (4.23)$$

Consequences for τ We have demonstrated that the average cluster size diverges when $p \to p_c$, which implies that the exponent γ must be positive. In turn, this implies that

$$\gamma > 0 \implies \frac{3 - \tau}{\sigma} > 0 \implies 3 > \tau \ . \qquad (4.24)$$

We have therefore found a first bound for τ: $\tau < 3$. As an exercise, you can check that this relation holds for the one-dimensional system and the Bethe lattice.

Density of Spanning Cluster

We may use a similar argument to find the behavior of $P(p)$ from the cluster number density, which will give us further scaling relations between exponents and another bound on the exponent τ.

Relation Between $P(p)$ and $n(s, p)$ We recall the general relation

$$\sum_s sn(s, p) + P(p) = p \ . \qquad (4.25)$$

This equation expresses that a site picked at random is occupied with probability p (right hand side), and that this site must either be in a finite cluster, with a probability corresponding to the sum $\sum_s sn(s, p)$, or in the infinite cluster with probability $P(p)$. We can therefore find $P(p)$ from

$$P(p) = p - \sum_s sn(s, p) . \tag{4.26}$$

by calculating the sum on the right hand side.

Finding the Sum Using the Scaling Ansatz We can find the sum over $sn(s, p)$ when p is close to p_c by transforming the sum to an integral and inserting the scaling ansatz $n(s, p) = s^{-\tau} F(s/s_\xi)$:

$$\sum_{s=1}^{\infty} sn(s, p) \simeq \sum_{s=1}^{\infty} ss^{-\tau} F(s/s_\xi) . \tag{4.27}$$

Again, we approximate the sum with an integral over s:

$$\sum_{s=1}^{\infty} s^{1-\tau} F(s/s_\xi) \simeq \int_{1}^{\infty} s^{1-\tau} F(s/s_\xi) ds . \tag{4.28}$$

Here, $F(s/s_\xi)$ is approximately a constant when $s < s_\xi$ and goes very rapidly to zero when $s > s_\xi$, so we integrate up to s_ξ assuming that $F(s/s_\xi)$ is a constant C up to s_ξ, giving

$$\int_{1}^{\infty} s^{1-\tau} F(s/s_\xi) ds \simeq \int_{1}^{s_\xi} Cs^{1-\tau} ds = c_1 + c_2 s_\xi^{2-\tau} . \tag{4.29}$$

We insert this back into the expression for $P(p)$ in (4.26) getting:

$$P(p) = p - \sum_s sn(s, p) \simeq p - c_1 - c_2 s_\xi^{2-\tau} . \tag{4.30}$$

Consequences for τ First, we realize that $P(p)$ cannot diverge when $p \to p_c$. Since s_ξ diverges, this means that the exponent $2 - \tau$ must be smaller than or equal to zero, otherwise $P(p)$ will diverge. This gives us a new bound for τ:

$$2 - \tau \leq 0 \Rightarrow 2 \leq \tau . \tag{4.31}$$

This means that τ is bounded by 2 and 3: $2 \leq \tau < 3$. This is an impressive result from the scaling ansatz.

Relating the Exponents β and τ We can rewrite the expression in (4.30) for $P(p)$ and insert $s_\xi = s_0 |p - p_c|^{-1/\sigma}$, getting:

$$P(p) \simeq p - c_1 - c_2 s_\xi^{2-\tau} \simeq (p - p_c) + c_2 \left(|p - p_c|^{-1/\sigma} \right)^{2-\tau} \tag{4.32}$$

We realize that when $p \to p_c$ the linear term $(p - p_c)$ will be smaller than the term $|p - p_c|^{(\tau-2)/\sigma}$. And we remember that $P(p) \propto (p - p_c)^\beta$. This gives us a new scaling relation for β:

$$\beta = \frac{\tau - 2}{\sigma} . \tag{4.33}$$

We have therefore again demonstrated the power of the scaling ansatz by both calculating bounds for τ and by finding relations between the scaling exponents.

4.3 Percolation Thresholds

While the exponents are universal and independent of the details of the lattice but dependent on the dimensionality, the percolation threshold, p_c, depends on details of the system such as the lattice type and the type of percolation. We typically discern between *site percolation*, where neighboring sites on a lattice are connected if present, and *bond percolation*, where the presence of bonds between the sites determines the connectivity. Table 4.2 provides basic values for the percolation thresholds. These results have been measured with code in this book and therefore have limited precision. You can find an updated set of percolation threshold for various models on the Wikipedia page for percolation at https://en.wikipedia.org/wiki/Percolation_threshold.

Table 4.2 Percolation thresholds for various models

Lattice type	Site	Bond
$d = 1$	1	1
$d = 2$		
Square	0.5927	1/2
Triangular	1/2	0.34
$d = 3$		
Cubic	0.3	0.25

Exercises

Exercise 4.1 (Alternative Way to Analyze Percolation Clusters) In this exercise we will use python to generate and visualize percolation clusters. We generate a $L \times L$ matrix of random numbers, and will examine clusters for a occupation probability p.

We generate the percolation matrix consisting of occupied (1) and unoccupied (0) sites, using

```
import numpy as np
import matplotlib.pyplot as plt
from scipy.ndimage import measurements
L = 100
r = np.random.rand(L,L)
p = 0.6
z = r<p # This generates the binary array
lw, num = measurements.label(z)
```

We have then produced the array `lw` that contains labels for each of the connected clusters.

(a) Familiarize yourself with labeling by looking at `lw`, and by studying the second example in the python help system on the image analysis toolbox.

We can examine the array directly by mapping the labels onto a color-map, using `imshow`.

```
plt.imshow(lw)
```

We can extract information about the labeled image using `measurements`, for example, we can extract an array of the areas of the clusters using

```
labelList = np.arange(lw.max() + 1)
area = measurements.sum(z, lw, labelList)
```

You can also extract information about the clusters using the `skimage.measure` module. This provides a powerful set of tools that can be used to characterize the clusters in the system. For example, you can determine if a system is percolating by looking at the extent of a cluster. If the extent in any direction is equal to L, then the cluster is spanning the system. We can use this to find the area of the spanning cluster or to mark if there is a spanning cluster:

```
import skimage
props = skimage.measure.regionprops(lw)
spanning = False
for prop in props:
    if (prop.bbox[2]-prop.bbox[0]==L or
        prop.bbox[3]-prop.bbox[1]==L):
        # This cluster is percolating
        area = prop.area
        spanning = True
        break
```

(b) Using these features, write a program to calculate $P(p, L)$ for various p for the two-dimensional system.

(c) How robust is your algorithm to changes in boundary conditions? Could you do a rectangular grid where $L_x \gg L_y$? Could you do a more complicated set of boundaries? Can you think of a simple method to ensure that you can calculate P for any boundary geometry?

Exercise 4.2 (Finding $\Pi(p, L)$ and $P(p, L)$)

(a) Write a program to find $P(p, L)$ and $\Pi(p, L)$ for $L = 2, 4, 8, 16, 32, 64, 128$. Comment on the number of samples you need to make to get a good estimate for P and Π.

(b) Test the program for small L by comparing with the exact results from above. Comment on the results?

Exercise 4.3 (Determining β) We know that when $p > p_c$, the probability $P(p, L)$ for a given site to belong to the percolation cluster, has the form

$$P(p, L) \sim (p - p_c)^\beta . \tag{4.34}$$

Use the data from above to find an expression for β. For this you may need that $p_c = 0.59275$.

Exercise 4.4 (Determining the Exponent of Power-Law Distributions) In this exercise you will build tools to analyse power-law type probability densities.

Generate the following set of data-points in python:

```
import numpy as np
z = np.random.rand(int(1e6))**(-3+1)
```

Your task is to determine the distribution function $f_Z(z)$ for this distribution.

Hint The distribution is on the form $f(u) \propto u^\alpha$.

(a) Find the cumulative distribution, that is, $P(Z > z)$. You can then find the actual distribution from

$$f_Z(z) = \frac{dP(Z > z)}{dz} . \tag{4.35}$$

(b) Generate a method to do logarithmic binning in python. That is, you estimate the density by doing a histogram with bin-sizes that increase exponentially in size.

Hint Remember to divide by the correct bin-size.

Exercise 4.5 (Cluster Number Density $n(s, p)$**)** We will generate the cluster number density $n(s, p)$ from the two-dimensional data-set.

Hint 1 The cluster sizes are extracted using `area = measurements.sum(z, lw, labelList)` as described in a previous exercise.

Hint 2 Remember to remove the percolating cluster.

Hint 3 Use logarithmic binning.

(a) Estimate $n(s, p)$ for a sequence of p values approaching $p_c = 0.59275$ from above and below.
(b) Estimate $n(s, p_c; L)$ for $L = 2^k$ for $k = 4, \ldots, 9$. Use this plot to estimate τ.
(c) Can you estimate the scaling of $s_\xi \sim |p - p_c|^{-1/\sigma}$ using this data-set?

Hint 1 Use $n(s, p)/n(s, p_c) = F(s/s_\xi) = 0.5$ as the definition of s_ξ.

Exercise 4.6 (Average Cluster Size)

(a) Find the average (finite) cluster size $S(p)$ for p close to p_c, for p above and below p_c.
(b) Determine the scaling exponent $S(p) \sim |p - p_c|^{-\gamma}$.
(c) In what ways can you generate $S^{(k)}(p)$? What do you think is the best way?

Geometry of Clusters

5

We have seen how we can characterize clusters by their mass, s. As p approaches p_c, the typical cluster size s increases as well as the characteristic cluster diameter. In this chapter we will discuss the geometry of clusters, and by geometry we will mean how the number of sites in a cluster is related to the linear size of the cluster. We will introduce several measures to characterize the spatial extent, the characteristic radius R_s, of clusters of size s. We will measure R_s to motivate that it is proportional to s^1/D, where D is a new exponent characterizing the dimension of clusters. We will demonstrate that the percolation system is characterized by two lengths, the system size L and a characteristic cluster size ξ, and that the system shows fractal, self-similar scaling when the characteristic length diverges. We develop scaling theories for $P(s, L)$ for $p > p_c$ and lay the foundations for a geometrical understanding and description of the spanning cluster.

5.1   Geometry of Finite Clusters

We have so far studied the clusters in our model porous material, the percolation system, through the distribution of cluster sizes, $n(s, p)$, and properties that can be found from $n(s, p)$, such as the average cluster size, S and the characteristic cluster size, s_ξ. However, clusters with the same mass, s, can have very different shapes. Figure 5.1 illustrates three clusters all with $s = 20$ sites. Notice that the linear and the compact clusters are unlikely, but possible realizations. How can we characterize the spatial extent of these clusters?

There are many ways to define the extent of a cluster. We could, for example, define the maximum distance between any two points in a cluster i ($R_{\max,i}$) to be the extent of the cluster, or we could use the average distance between two points in the cluster. However, it is common to use the standard deviation of the position of the sites in a cluster, which we recognize as the radius of gyration of a cluster. The radius of gyration R_i for a cluster i of size s_i with sites at $\mathbf{r}_j$ for $j = 1, \ldots, s_i$, is

© The Author(s) 2024
A. Malthe-Sørenssen, *Percolation Theory Using Python*, Lecture Notes
in Physics 1029, https://doi.org/10.1007/978-3-031-59900-2_5

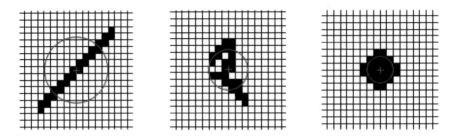

Fig. 5.1 Illustrations of three clusters all with $s = 24$. The red circle illustrates the radius of gyration for the clusters

defined as

$$R_i^2 = \frac{1}{s_i} \sum_{j=1}^{s_i} \left(\mathbf{r}_j - \mathbf{r}_{cm,i} \right)^q , \tag{5.1}$$

where $\mathbf{r}_{cm,j}$ is the center of mass of cluster i. The values for R_i of the clusters in Fig. 5.1 are illustrated by circles.

As we see from Fig. 5.1, clusters of the same size s can have different radii. How can we then find a characteristic size for a given cluster size s? We find that by averaging R_i^2 over all clusters of the same size s:

$$R_s^2 = \langle R_i^2 \rangle_i . \tag{5.2}$$

Let us see how this can be done analytically in one dimension and numerically in two dimensions.

Analytical Results in One Dimension

We can use the one-dimensional percolation system to gain insight into how we expect R_s to depend on s. In one dimension, a cluster of size s can only be realized in one way, as a line of length s. If the cluster runs from 1 to s, the center of mass is at $s/2$, and the sum over all sites runs from 1 to s:

$$R_s^2 = \frac{1}{s} \sum_{i=1}^{s} (i - s/2)^2 , \tag{5.3}$$

where we assume that s is so large that we only need to address the leading term in s, and that we do not have to treat even and odd s separately. This can be expanded to

$$R_s^2 = \frac{1}{s}[\sum_{i=1}^{s} i^2 - is + \frac{s^2}{4}] = \frac{1}{s}\left[\frac{s(s+1)(2s+1)}{6} - s\frac{s(s+1)}{2} + s\frac{s^2}{4}\right] \propto s^2,$$

(5.4)

where we have used that $\sum_{i=1}^{s} s^2 = s(s+1)(2s+1)/6$ and $\sum_{i=1}^{s} s = s(s+1)/2$ and where we only have kept the leading term in s. This shows that $R_s^2 \propto s^2$, which means that $s \propto R_s$ in one dimension. This is indeed what we expected. The extent of the cluster is proportional to s because the cluster is s sites long.

Numerical Results in Two Dimensions

For the one-dimensional system we found that $s \propto R_s$. How does this generalize to higher dimensions? We start by measuring the behavior for a given value of p for a finite system of size L. Our strategy is: (i) to generate clusters on a $L \times L$ lattice; (ii) for each cluster, i, of size s_i, we will find the center of mass and the radius of gyration, R_i^2; and (iii) for each value of s we will find the average radius, R_s^2, by a linear average. For larger values of s we will collect the data in bins, using the logaritmic binning approach we developed to measure $n(s, p)$.

Developing a Function to Measure R_s   First, we introduce a function to calculate the radius of gyration of all the clusters in a lattice. This is done in two steps, first we find the center of mass of all clusters, and then we find the radius of gyration. The center of mass for a cluster i with sites at $\mathbf{r}_{i,j}$ for $j = 1, \ldots, s_i$, is

$$\mathbf{r}_{cm,i} = \frac{1}{s_i} \sum_{j=1}^{s_i} \mathbf{r}_{i,j},$$

(5.5)

We generate a lattice, ensure that each cluster are marked with the index of the cluster, and find the center of mass cm using a built-in command:

```
L = 400
p = 0.58
z = np.random.rand(L,L)
m = z<p
lw, num = measurements.label(m)
cm = measurements.center_of_mass(m, lw, labelList)
```

Second, we calculate the radius gyration by running through all the sites ix,iy in the lattice. For each site, we find the index i of the cluster that it belongs to from i = lw[ix,iy]. If the site belongs to a cluster, that is if i>0, we add the sum of the square of the distance from the site to the center of mass to the radius of gyration for cluster i:

```
dr = np.array([ix,iy])-cm[i]
rad2[i] = rad2[2] + np.dot(dr,dr)
```

After running through all the site, we divide by the mass (size) s_i of each cluster to find the radius of gyration according to the formula

$$R_i^2 = \frac{1}{s_i} \sum_{j=1}^{s_i} \left(\mathbf{r}_{i,j} - \mathbf{r}_{cm,i} \right)^2 \ , \tag{5.6}$$

This is implemented in the following function:

```
import numpy as np
import matplotlib.pyplot as plt
from scipy.ndimage import measurements

def radiusofgyration(m,lw,L):
    labelList = np.arange(lw.max() + 1)
    area = measurements.sum(m, lw, labelList)
    cm = measurements.center_of_mass(m, lw, labelList)
    rad2 = np.zeros(int(lw.max()+1))
    for ix in range(L):
        for iy in range(L):
            ilw = lw[ix,iy];
            if (ilw>0):
                dr = np.array([ix,iy])-cm[ilw]
                dr2 = np.dot(dr,dr)
                rad2[ilw] = rad2[ilw] + dr2
    rad = np.sqrt(rad2/area)
    return area,cm,rad2

M = 20    # Nr of samples
L = 400   # System size
p = 0.58 # p-value
allr2 = np.array([])
allarea = np.array([])
for i in range(M):
    z = np.random.rand(L,L)
    m = z<p
    lw, num = measurements.label(m)
    area,rcm,rad2 = radiusofgyration(m,lw,L)
    allr2 = np.append(allr2,rad2)
    allarea = np.append(allarea,area)

plt.loglog(allarea,allr2,'k.')
```

We use this function to calculate the average radius of gyration for each cluster size s for M different lattice realizations, and plot the results using the following script:

```
M = 20    # Nr of samples
L = 400   # System size
p = 0.58 # p-value
allr2 = np.array([])
allarea = np.array([])
for i in range(M):
    z = np.random.rand(L,L)
    m = z<p
    lw, num = measurements.label(m)
    area,rcm,rad2 = radiusofgyration(m,lw,L)
    allr2 = np.append(allr2,rad2)
    allarea = np.append(allarea,area)
plt.loglog(allarea,allr2,'k.')
plt.xlabel("$s$")
plt.ylabel("$R_s^2$")
```

Scaling Behavior of the Radius of Gyration The resulting plots for several values of p are shown in Fig. 5.2. We see that there is an approximately linear relation between R_s^2 and s in this double-logarithmic plot, which indicates that there is a power-law relationship between the two:

$$R_s^2 \propto s^x \ . \tag{5.7}$$

How can we interpret this relation? Equation (5.7) relates the radius R_s and the area (or mass) of the cluster. We are more used to the inverse relation:

$$s \propto R_s^D \ , \tag{5.8}$$

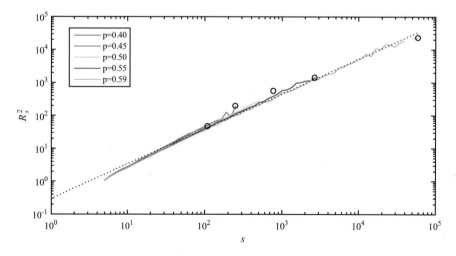

Fig. 5.2 Plot of R_s^2 as function of s for simulations on two-dimensional systems with $L = 400$. The largest cluster for each value of p is illustrated by a circle. The dotted line shows the curve $R_s^2 \propto s^{2/D}$ for $D = 1.89$

where $D = 2/x$ is the exponent relating the radius to the mass of a cluster. This corresponds to our intuition from geometry. We know that for a cube of size L, the mass (or volume) of the cube is $M = L^3$. For a square of length L, the mass (or area) is $M = L^2$, and similarly for a circle $M = \pi R^2$, where R is the radius of the circle. For a line of length L, the mass is $M = L^1$. We see a general trend, $M \propto R^d$, where R is a characteristic length for the object, and d describes the dimensionality of the object. If we extend this intuition to the relation in (5.8), which is an observation based on Fig. 5.2, we see that we may interpret D as the dimension of the cluster. However, the value of D is not an integer. We have indicated the value of $D = 1.89$ with a dotted line in Fig. 5.2. This non-integer value of D may seem strange, but it is fully possible, mathematically, to have non-integer dimensions. This is a feature frequently found in fractal structures, and percolation clusters as p approaches p_c are indeed good examples of self-similar fractals. We will return to this aspect of the geometry of the percolation system in Sect. 5.3.

Characteristic Cluster Radius The largest cluster and its corresponding radius of gyration is indicated by a circle for each p value in Fig. 5.2. We see that as p approaches p_c, both the mass and the radius of the largest cluster increases. Indeed, this corresponds to the observation we have previously made for the characteristic cluster size, s_ξ. We may define a corresponding *characteristic cluster radius*, R_{s_ξ} through

$$s_\xi \propto R_{s_\xi}^D \quad \Longleftrightarrow \quad R_{s_\xi} \propto s_\xi^{1/D} . \tag{5.9}$$

This length is a characteristic length for the system for a given value of p, corresponding to the largest cluster size or the typical cluster size in the system. In Sect. 5.2 we see how we can relate this length directly to the cluster size distribution.

Scaling Behavior in Two Dimensions

We have already found that the characteristic cluster size s_ξ diverges as a power law as p approaches p_c:

$$s_\xi \simeq s_0 (p - p_c)^{-1/\sigma} , \tag{5.10}$$

when $p < p_c$. The behavior is similar when $p > p_c$, but the prefactor s_0 may be different. How does R_{s_ξ} behave when p approaches p_c? We can find this by combining the scaling relations for s_ξ and R_{s_ξ} from (5.9):

$$R_{s_\xi} \propto s_\xi^{1/D} \propto \left((p - p_c)^{-1/\sigma}\right)^{1/D} \propto (p - p_c)^{-1/\sigma D} , \tag{5.11}$$

where we introduce the symbol $\nu = 1/(\sigma D)$. For two-dimensional percolation, the exponent ν is a universal number, just like σ and D. By universal, we mean that it does not depend on details such as the lattice type or the connectivity of the lattice, although it does depend on the dimensionality of the system. We know the exact value of ν in two dimensions, $\nu = 4/3$.

The argument we have provided here is an example of a *scaling argument*. In these arguments we are only interested in the exponent in the scaling relation, which gives us the functional form, and not in the values of the prefactors.

5.2 Characteristic Cluster Size

We have now defined the characteristic size of a cluster of size s through R_s. In addition, we introduced a characteristic cluster length R_{s_ξ}, which characterizes the whole system and not only clusters of a particular size s. However, there are several ways we can introduce a length scale to describe the typical cluster size in a system. Here, we will introduce two such measures, the average radius of gyration R and the characteristic length ξ.

Average Radius of Gyration

We have now defined the characteristic length R_{s_ξ} through the definition of the characteristic cluster size, s_ξ, and the scaling relation $s \propto R_s^D$. However, it may be more natural to define the characteristic length of the system as the *average* radius and not the *cut-off* radius. We introduced the radius of gyration for clusters of size s by averaging the radius of gyration R_i over all clusters i of size s:

$$R_s^2 = \langle R_i^2 \rangle_i \, , \tag{5.12}$$

This gives us the radius of gyration R_s, which we found to scale with cluster mass s as $s \propto R_s^D$.

Introducing an Average Cluster Radius For the cluster sizes, we introduced an average cluster size S, which is

$$S = \frac{1}{Z_S} \sum_s s \, sn(s, p) \, , \ Z_S = \sum_s sn(s, p) \, . \tag{5.13}$$

We can also similarly introduce an *average radius of gyration, R,* by averaging R_s over all cluster sizes:

$$R = \frac{1}{Z_R} \sum_s R_s^2 s^k sn(s, p) \, , \ Z_R = \sum_s s^k sn(s, p) \, . \tag{5.14}$$

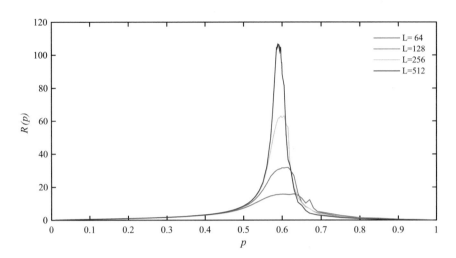

Fig. 5.3 A plot of the average radius of gyration R as a function of p and L. We observe that R increases as $p \to p_c$, but is limited in magnitude by the finite system size L

Here, we have purposely introduced an unknown exponent k. We are to some extent free to choose this exponent, although the average needs to be finite, and the exponent will determine how small and large clusters are weighed in the sum. A natural choice may be to choose $k = 1$ so that we get terms $R_s^2 s^2 n(s, p)$ in the sum. However, the results we present here will not change in any significant way, expect for different prefactors to the scaling relations, if you choose a larger value of k. Using $k = 1$, we define the average radius of gyration to be

$$R = \frac{1}{Z_R} \sum_s R_s^2 s^2 n(s, p) \, , \quad Z_R = \sum_s s^2 n(s, p) \, , \tag{5.15}$$

where we notice that the normalization sum Z_R is the average cluster size, S, $Z_R = S$. Figure 5.3 shows a plot of the average R as a function of p for various systems sizes L. We see that R diverges as p approaches p_c. How can we develop a theory for this behavior?

A Scaling Form for R   We know that the cluster number density $n(s, p)$ has the approximate scaling form

$$n(s, p) = s^{-\tau} F\left(s/s_\xi\right) \, , \quad s_\xi \propto |p - p_c|^{-1/\sigma} \, . \tag{5.16}$$

We can use this to calculate the average radius of gyration, R, when p is close to p_c. We find the scaling behavior of the average radius of gyration by replacing the sums over s with integrals over s:

$$R^2 = \frac{\sum_s R_s^2 s^2 n(s, p)}{\sum_s s^2 n(s, p)} = \frac{\int_1^\infty R_s^2 s^{2-\tau} F(s/s_\xi)\, ds}{\int_1^\infty s^{2-\tau} F(s/s_\xi)\, ds} \tag{5.17}$$

$$\propto \frac{\int_1^\infty s^{2/D} s^{2-\tau} F(s/s_\xi)\, ds}{\int_1^\infty s^{2-\tau} F(s/s_\xi)\, ds}, \tag{5.18}$$

where we have inserted $R_s^2 \propto s^{2/D}$. While this expression only is valid when $s < s_\xi$, we can apply it here since $F(s/s_\xi)$ goes rapidly to zero when $s > s_\xi$, and therefore only the $s < s_\xi$ values will contribute significantly to the integral. We change variables to $u = s/s_\xi$, getting:

$$R^2 \propto \frac{s_\xi^{2/D+3-\tau} \int_{1/s_\xi}^\infty u^{2/D+2-\tau} F(u)\, du}{s_\xi^{3-\tau} \int_{1/s_\xi}^\infty u^{2-\tau} F(u)\, du} \tag{5.19}$$

$$\propto s_\xi^{2/D} \frac{\int_0^\infty u^{2/D+2-\tau} F(u)\, du}{\int_0^\infty u^{2-\tau} F(u)\, du} \propto s_\xi^{2/D}, \tag{5.20}$$

where the lower limit $1/s_\xi$ goes to zero for large s_ξ and the two integrals over $F(u)$ simply are numbers and therefore have been included in the constant of proportionality.

The Characteristic Lengths Are Proportional   This shows that $R^2 \propto s_\xi^{2/D}$. We found above that $R_{s_\xi} \propto s_\xi^{2/D}$. Therefore, $R \propto R_{s_\xi}$! These two characteristic lengths therefore have the same behavior. They are only different by a constant of proportionality, $R = c\, R_{s_\xi}$. We can therefore use either length to characterize the system—they are effectively the same up to a prefactor. This is not only true for these two lengths, but all lengths have the same asymptotic scaling behavior close to p_c. For example, Fig. 5.4 illustrates the radius of gyration of the largest cluster with a circle and the average radius of gyration, R, by the length of the side of the square. As p increases, both the maximum cluster size and the average cluster size increases in concert.

Correlation Length

We can also define the typical size of a cluster from the correlation function. We recall that the correlation function $g(r, p)$ is the probability that an occupied site i at $\mathbf{r}_i$ is connected to a site j at $\mathbf{r}_j$, where $\mathbf{r} = \mathbf{r}_j - \mathbf{r}_i$ and $r = |\mathbf{r}|$. The correlation function is only a function of the relative position of the two sites, $\mathbf{r}$, which we usually only write as r, because we assume that the correlation function is isotropic. We define the correlation length, ξ, as the average squared distance between two

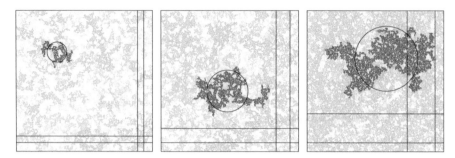

Fig. 5.4 Illustration of the largest cluster in 512×512 systems for $p = 0.55$, $p = 0.57$, and $p = 0.59$. The circles illustrate the radius of gyration of the largest cluster, and the boxes show the size of the average radius of gyration, $R = \langle R_s \rangle$. We observe that both lengths increase approximately proportionally as p approaches p_c

connected sites:

$$\xi^2 = \frac{\sum_{\mathbf{r}} r^2 g(r; p)}{\sum_{\mathbf{r}} g(r; p)} \; . \tag{5.21}$$

where the sum is over all relative positions $\mathbf{r}$, that is, over all space. The denominator is a normalization sum, which corresponds to the average cluster size, S. This length is called the *correlation length*. However, to gain insight into this length, we will first address the correlation function, its scaling behavior and its relation to the average cluster size S.

One-Dimensional System   In Chap. 2 we found that for a one-dimensional system, the correlation function $g(r; p)$ is

$$g(r) = p^r = e^{-r/\xi} \; , \tag{5.22}$$

where $\xi = -\frac{1}{\ln p} =\simeq 1/(1 - p_c)$ is called the *correlation length*. The correlation length diverges as $p \to p_c = 1$ as $\xi \simeq (1 - p_c)^{-\nu}$, where $\nu = 1$.

We can generalize this behavior by writing the correlation function in a more general scaling form for the one-dimensional system

$$g(r; p) = r^0 f(r/\xi) \; , \tag{5.23}$$

where $f(u)$ is a function that decays rapidly when u is larger than 1. We will assume that this behavior can be generalized to higher dimension. That is, we expect the correlation function to decay rapidly beyond a length, ξ, that corresponds to the typical extent of clusters in the system.

Measuring the Correlation Function   For a two- or three-dimensional system, we cannot find the exact form of the correlation function, like we could in one dimension. However, we can still measure it from our simulations, although such measurements typically are computationally intensive. How can we measure it? We can loop through all sites i and j and find their distance r_{ij}. We estimate the probability for two sites at a distance r_{ij} to be connected by counting how many of the sites that are a distance r_{ij} apart that are connected and compare it to how many sites in total that are a distance r_{ij} apart. This is done through the following program, which returns the correlation function $g(r)$ estimated for a lattice lw which contains the cluster indexes for each site, similar to what is returned by the lw, num = measurements.label(m) command. We write a subroutine perccorrfunc to find the correlation function for a given lattice lw, and then we use this function to find the correlation function for several values of p and L:

```python
import numpy as np
import matplotlib.pyplot as plt
from scipy.ndimage import measurements
from numba import jit
@jit
def perccorrfunc(m,lw,L):
    r = np.arange(2*L) # Positions
    pr = np.zeros(2*L)    # Correlation function
    npr = np.zeros(2*L)  # Nr of elements
    for ix1 in range(L):
        for iy1 in range(L):
            lw1 = lw[ix1,iy1]
            if (lw1>0):
                for ix2 in range(L):
                    for iy2 in range(L):
                        lw2 = lw[ix2,iy2]
                        if (lw2>0):
                            dx = (ix2-ix1)
                            dy = (iy2-iy1)
                            rr = np.hypot(dx,dy)
                            # Find corresponding box
                            nr = int(np.ceil(rr)+1)
                            pr[nr] = pr[nr] + (lw1==lw2)
                            npr[nr] = npr[nr] + 1
    pr = pr/npr
    return r,pr

# Calculate correlation function
M = 20 # Nr of samples
L = 800 # System size
pp = [0.57,0.58,0.59] # p-value
lenpp = len(pp)
pr = np.zeros((2*L,lenpp),float)
rr = np.zeros((2*L,lenpp),float)
for i in range(M):
    print("i = ",i)
    z = np.random.rand(L,L)
    for ip in range(lenpp):
        p = pp[ip]
```

```
            m = z<p
            lw, num = measurements.label(m)
            r,g = perccorrfunc(m,lw,L)
            pr[:,ip] = pr[:,ip] + g
            rr[:,ip] = rr[:,ip] + r
    pr = pr/M
    r = r/M

    # Plot data - linearly binned
    for ip in range(lenpp):
        plt.loglog(rr[:,ip],pr[:,ip],'.',label="p="+str(pp[ip]))
    plt.legend()
```

Figure 5.5 shows the resulting plots of the correlation function $g(r; p)$ for various values of p for an $L = 400$ system. First, we notice that the scaling is rather poor. We will understand this as we develop a theory for $g(r; p)$ below. The plot shows that there also in two dimensions appear to be a cross-over length ξ, which we call the correlation length, beyond which the correlation function falls rapidly to zero. For $r < \xi$ the correlation function appears to approximately be a power-law. Based on our experience with percolation systems, we suggests the following functional form

$$g(r; p) = r^{-x} f(r/\xi) , \qquad (5.24)$$

where the cross-over function $f(u)$ falls rapidly to zero when $u > 1$ and is approximately constant when $u < 1$. We expect that as p approaches p_c,

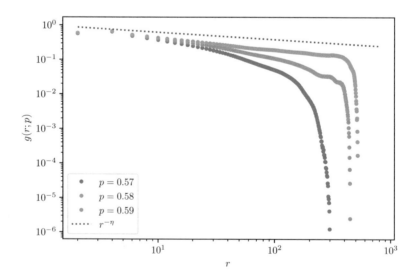

Fig. 5.5 A plot of $g(r; p)$ as a function of r for various values of p. When p approaches p_c the function approaches a power-law behavior $g(r) \propto r^{-\eta}$, which is illustrated by a dashed line with $\eta = 0.208$. Notice the rapid cross-over for large r, which occurs at a characteristic length ξ

the correlation length ξ grows to infinity, and the correlation function $g(r; p_c)$ approaches a power-law r^{-x} for all values of r up to a length limited by the system size L.

Relating the Correlation Function to $S(p)$ Based on these observations, we are motivated to develop a theory for the behavior of the correlation function. Our plan is first to relate the correlation function $g(r; p)$ to the average cluster size $S(p)$ and then use what we know about the behavior of $S(p)$ to determine the behavior of $g(r; p)$.

How can we relate $g(r; p)$ to $S(p)$? We notice that $S(p)$ is the average number of sites in a cluster, that is, the average number of sites connected to a given occupied site. We can therefore find $S(p)$ by summing the probability for a site to be connected for all possible relative positions $\mathbf{r}$:

$$S(p) = \sum_{\mathbf{r}} g(r; p) . \tag{5.25}$$

We approximate the sum by an integral:

$$S(p) = \sum_{\mathbf{r}} g(r; p) \simeq \int g(r) \, dr^d , \tag{5.26}$$

where the integral is over a volume in space corresponding to all relative positions $\mathbf{r}$. We change integration variables to the radial distance r and the solid angle Ω

$$S(p) \simeq \int g(r) dr^d = \int \int g(r) r^{d-1} \, dr \, d\Omega , \tag{5.27}$$

where the integral is from $r = 0$ to $r \to \infty$ and over all solid angles Ω.

Average Cluster Size at p_c We know that when $p \to p_c$, then the average cluster size, S, diverges. At $p = p_c$, the scaling form for the correlation function in (5.24) is $g(r; p) \propto r^{-x}$. The condition that the integral in (5.27) must diverge therefore provides bounds for the exponent x:

$$S(p_c) \simeq \int \int r^{-x} r^{d-1} \, dr \, d\Omega = c \int_1^\infty r^{-x+d-1} dr . \tag{5.28}$$

This integral diverges when $-x + d > 0$, that is, for $x < d$. It is common to introduce an exponent η so that $x = (d - 2 + \eta)$. The condition for $S(p_c)$ to diverge is then $2 - \eta > 0$. The correlation function at p_c is then:

$$g(r; p_c) \propto r^{-(d-2)-\eta} . \tag{5.29}$$

This is consistent with what we found for the one-dimensional system where $x = 0$ and $\eta = 1$.

Finding $S(p)$ for p below p_c We use the scaling form of $g(r; p)$ in (5.24) to calculate the integral in (5.27) for $p < p_c$:

$$S(p) \simeq \int \int r^{-(d-2)-\eta} f(\frac{r}{\xi}) r^{d-1} \, dr \, d\Omega = c \int_1^\infty r^{1-\eta} f(\frac{r}{\xi}) \, dr . \tag{5.30}$$

We change variables by introducing $u = r/\xi$ and realize that $f(u)$ is approximately a constant for $u < 1$ and zero for $u > 1$. We therefore use 1 as the upper limit for the integral since $f(u)$ rapidly goes to zero beyond $u = 1$.

$$S(p) \simeq c \int_{1/\xi}^\infty \xi^{2-\eta} u^{1-\eta} f(u) \, du = c\xi^{2-\eta} \int_{1/\xi}^1 u^{1-\eta} \, du = c\xi^{2-\eta}(1 - \xi^{-(2-\eta)}) . \tag{5.31}$$

Because $2 - \eta > 0$, we see that $\xi^{-(2-\eta)}$ approaches zero as p approaches p_c and ξ grows. The right-hand term is therefore approximately 1, and we get:

$$S(p) \propto \xi^{2-\eta} . \tag{5.32}$$

We also know that $S(p) \propto |p - p_c|^{-\gamma}$ so that

$$S(p) \propto \xi^{2-\eta} \propto |p-p_c|^{-\gamma} \quad \Rightarrow \quad \xi \propto |p-p_c|^{-\gamma/(2-\eta)} = |p-p_c|^{-\nu} , \tag{5.33}$$

where we have introduced a new critical exponent, ν, and related it to other exponents through $\nu = \gamma/(2 - \eta)$. For percolation in two dimensions, $\nu = 4/3$, whereas in three dimensions it is $\nu = 0.9$.

Finding the Correlation Length from $n(s, p)$ The correlation length ξ can be found from the correlation function by

$$\xi^2 = \frac{\sum_r r^2 g(r)}{\sum_r g(r)} , \tag{5.34}$$

where $\sum_r g(r) = S(p)$. You can check this by inserting $g(r) = r^{-(d-2)-\eta} f(r/\xi)$ and calculating the integral.

We recall that the average radius of gyration is

$$R^2 = \frac{\sum_s R_s^2 s^2 n(s, p)}{\sum_s s^2 n(s, p)} , \tag{5.35}$$

averaged over all clusters of size i. Here, $\sum_s s^2 n(s, p) = S(p)$. Notice the similarity between these two definitions. This indicates that the correlation length can be defined in terms on $n(s, p)$. The common definition is

$$\xi^2 = \frac{\sum_s 2R_s^2 s^2 n(s, p)}{\sum_s s^2 n(s, p)} = 2R^2 . \tag{5.36}$$

The important aspect is that the two lengths ξ and R are proportional to each other. And we already round that $R \propto R_{s_\xi}$. This means that they all have the same scaling behavior as p approaches p_c. This means that

$$\xi \propto |p - p_c|^{-\gamma/(2-\eta)} \propto |p - p_c|^{-\nu} \propto R_{s_\xi} \propto |p - p_c|^{-1/\sigma D} . \tag{5.37}$$

This gives us the scaling relation for η: $\eta = 2 - \gamma \sigma D$.

Correlation Length All the lengths R, R_{s_ξ} and ξ has the same scaling behavior.

The correlation length ξ scales as

$$\xi \propto |p - p_c|^{-\nu} \quad \text{when} \quad p \to p_c , \tag{5.38}$$

with the exponent $\nu = 1/(\sigma D) = \gamma/(2 - \eta)$. For a two-dimensional system, $\nu_{2d} = 4/3$, and for a three-dimensional system, $\nu_{3d} = 0.9$.

The Characteristic Length ξ and System Size L What happens to ξ in a finite system as p approaches p_c? Figure 5.3 shows a plot of $R(p) \propto \xi(p)$ for $L = 100$, 200, and 400 in two dimensions. Notice that the R does not diverge as p approaches p_c. Instead, it reaches a plateau, the height of which increases with system size L. This is not surprising, since we cannot observe clusters that are larger than the system size L. Figure 5.6 illustrates $\xi \propto |p - p_c|^\nu$ as we would expect it in an infinite system. However, for a finite system, the curve for $\xi(p)$ is cut off at a length proportional to L. This means that for p in some region around p_c, as illustrated in Fig. 5.6, we have that $\xi > L$. In this range we cannot determine if the system is at p_c or not, because the system is not large enough for us to make this distinction.

If we study a system of size $L \ll \xi$, we will typically observe a cluster that spans the system, since the typical cluster size, ξ, is larger than the system size. We are therefore not able to determine if we observe a spanning cluster because we are at p_c or only because we are sufficiently close to p_c. We will start to observe a spanning cluster when $\xi \simeq L$, which corresponds to

$$\xi_-(p_c - p)^{-\nu} = \xi \simeq L , \tag{5.39}$$

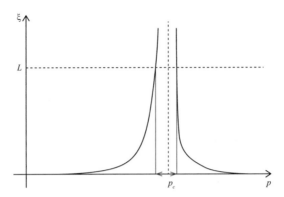

Fig. 5.6 Illustration of the behavior of ξ when p approaches p_c. In a finite system of size L, the system will be percolating for all p in the range indicated by the arrows

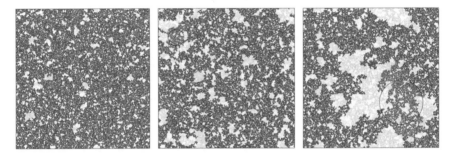

Fig. 5.7 Illustration of the largest cluster in 512×512 systems with $p > p_c$, for $p = 0.593$, $p = 0.596$, and $p = 0.610$. The circles illustrate the radius of gyration of the largest cluster. We observe that the radius of gyration increases as p approaches p_c

and therefore that

$$(p_c - p) \simeq (L/\xi_-)^{-(1/\nu)} , \qquad (5.40)$$

when $p < p_c$, and a similar expression for $p > p_c$. This means that when we observe spanning we can only be sure that p is within a certain range of p_c:

$$|p - p_c| = cL^{-1/\nu} . \qquad (5.41)$$

The correlation length ξ is therefore the natural length to characterize the system. At distances smaller than ξ, the system behaves as if it is at $p = p_c$. However, at distances much larger than ξ, the system is essentially homogeneous. As we can observe in Fig. 5.7 the system becomes more and more homogeneous when p goes away from p_c. We will now address this feature in more detail when $p > p_c$.

5.3   Geometry of the Spanning Cluster

How can we develop a scaling theory for the spanning cluster, that is, a theory for how the mass of the spanning cluster depends on system size L and the characteristic cluster size, ξ? We know that as p is increased from below towards p_c, the characteristic cluster size ξ diverges. The mass of a characteristic cluster of size ξ is expected to follow the scaling relation $s_\xi \propto \xi^D$. For a given value of p, we can therefore choose the system size L to be equal to ξ, $L = \xi(p)$. In this case, a cluster of size ξ would correspond to a cluster of size L, and it would be a spanning cluster in this system. For this system of size $L = \xi$, we therefore expect the mass of the spanning cluster to be $M(p, L) \propto \xi^D \propto L^D$. This suggests that the mass of the spanning cluster in a system close to or at p_c depends on the system size L according to $M(p_c, L) \propto L^D$.

The Density of the Spanning Cluster The density, $P(p, L)$, of the spanning cluster at $p = p_c$ therefore has the following behavior:

$$P(p_c, L) = \frac{M(p_c, L)}{L^d} \propto L^D/L^d \propto L^{D-d} . \tag{5.42}$$

Because we know that $P(p_c, L)$ does not diverge when $L \to \infty$, we deduce that $D < d$. The value of D in two-dimensional percolation is $D = 91/48 \simeq 1.90$.

This implies that the density of the spanning cluster depends on the system size, L. Indeed, since $D < d$, we see that the density decreases with system size. This may initially seem surprising, since we may be used to thinking of density as a material property. However, we recognize this behavior from bodies that are embedded in dimensions higher than themselves, such as for a thin sheet or a thin rod embedded in three dimensions.

Embedded, Regular Bodies For example, consider a thin, flat sheet of thickness h, and dimensions $\mathscr{L} \times \mathscr{L}$, placed in a three-dimensional space. If we cut out a volume of size $L \times L \times L$, so that $L \ll \mathscr{L}$, the mass of the sheet inside that volume is

$$M = hL^2 , \tag{5.43}$$

which implies that the density of the sheet is

$$\rho = \frac{hL^2}{L^3} = hL^{-1} . \tag{5.44}$$

It is only in the case when we use a two-dimensional volume $L \times L$ with a third dimension of constant thickness H larger than h, that we recover a constant density ρ independent of system size.

Fig. 5.8 Illustration of three generations of the Sierpinski gasket starting from an equilateral triangle

Non-Integer, Fractal Dimensions We found that D was indeed smaller than d so that the density decreases with system size. However, D is also not an integer. How can we build intuition for non-integer dimensions of objects? First, let us be precise about what we mean with *dimension*.

> **Dimension** For an object with mass M and linear size L, we define the dimension of the object as D, if $M(L) = cL^D$, that is, if the mass is proportional to L to the power of D.

Self-Similar Deterministic Fractals To gain intuition about non-integer values for the dimension D, we will introduce a structure known as a deterministic fractal. A famous example is the Sierpinski gasket [32], which is defined iteratively. We start with a unit equilateral triangle as illustrated in Fig. 5.8. We divide the triangle into 4 identical triangles, and remove the center triangle. For each of the remaining triangles, we continue this process. The resulting set of points after infinitely many iterations is called the Sierpinski gasket. This set contains a hierarchy of holes. We also notice that the structure is identical under (a specific) dilatational rescaling. If we take one of the tree triangles generated in the first step and rescale it to fit on top of the initial triangle, we see that it reproduces the original identically. This structure is therefore a *self-similar fractal* in the limit of an infinite number of generations of iterations.

To find the dimensionality of such a structure we need to understand how the mass M depends on the length scale L of the structure. A common trick is to look at how the mass is rescaled between generations of iterations of the structure. Each time we generate a new iteration of the structure we increase the length scale by a factor of 2 and the mass by a factor of 3. This means that $L' = 2L$ and $M' = M(2L) = 3M(L)$. If we assume that $M(L) = cL^D$, we get that

$$M(2L) = c(2L)^D = 3M(L) = 3cL^D \quad \Rightarrow \quad 2^D = 3 , \tag{5.45}$$

where we take the logarithm on both sides, getting

$$D \ln 2 = \ln 3 \quad \Rightarrow \quad D = \frac{\ln 3}{\ln 2} \simeq 1.58 . \tag{5.46}$$

Thus, the dimension of this object is also not an integer, but lies between 1 and 2. This indicates that the Sierpinski gasket is an object that is somewhere between a plane and a line. It fills space less efficiently that a plane, but more efficiently than a line.

We have presented a robust method for finding the dimension of iteratively defined objects. However, the method also gives us the correct dimension for e.g. a cube: For a cube of size L, if we double the size of the cube, that is $L' = 2L$, the mass is increased by a factor of 8, $M' = M(2L) = 8M(L)$, which gives a dimension of $d = \ln 8 / \ln 2 = 3$.

Box Counting Typically, the mass dimensions of objects from experiments or simulations are measured by *box counting*. The sample is divided into regular boxes where the size of each side of the box is δ. The number of boxes, $N(\delta)$, that contain the cluster are counted as a function of δ. For a uniform mass we expect

$$N(\delta) = (\frac{L}{\delta})^d , \tag{5.47}$$

and for a fractal structure we expect

$$N(\delta) = (\frac{L}{\delta})^D , \tag{5.48}$$

We leave it as an exercise for the reader to address what happens when $\delta \to 1$, and when $\delta \to L$.

5.4 Spanning Cluster Above p_c

Let us now return to the discussion of the mass $M(p, L)$ of the spanning cluster for $p > p_c$ in a finite system of size L. The behavior of the percolation system for $p > p_c$ is illustrated in Fig. 5.7. We notice that the correlation length ξ diverges when p approaches p_c. At lengths larger than ξ, the system is effectively homogeneous because there are no holes significantly larger than ξ. For such systems, there are two effectively types of behavior, depending on whether L is larger than or smaller than the correlation length ξ.

Systems Effectively at p_c When $L \ll \xi$ and $p > p_c$, we are in the situation where we cannot discern a system at p from a system at p_c because the size of the holes (empty regions described by ξ when $p > p_c$) in the spanning cluster is much larger than the system size. We say that the system is effectively at p_c. When we look at

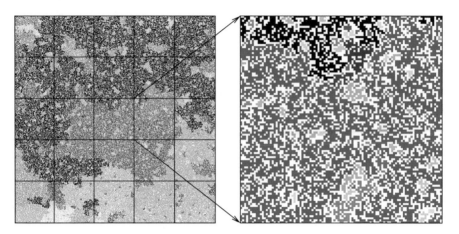

Fig. 5.9 Illustration of the spanning cluster in a 512×512 system at $p = 0.595 > p_c$. In this case, the correlation length is $\xi = 102$. The system is divided into regions of size ξ. Each such region has a mass $M(p, \xi) \propto \xi^D$, and there are $(L/\xi)^d \simeq 25$ such regions in the system

this system through a window of size L, that is, in a finite system size, we do not know if the spanning cluster in that system is an infinite cluster or just a cluster that is larger than the system size L. However, because the mass of finite clusters all scale as $s \propto R_s^D$ up to a length ξ that is much larger than L, the mass of the spanning cluster of size L will also follow this scaling relation:

$$M(p, L) \propto L^D \text{ when } L \ll \xi \ . \tag{5.49}$$

Systems Away from p_c In the other case, when $L \gg \xi$ and $p > p_c$, the typical size of a hole in the percolation cluster is ξ, as illustrated in Fig. 5.9. This means that on lengths much larger than ξ, the percolation cluster is effectively homogeneous. That is, it has holes up to a size ξ, but not much larger than that. It looks like a Swiss cheese with a finite size of the holes. We can therefore divide the $L \times L$ system into $(L/\xi)^d$ regions of size ξ. In each such region of size $\ell = \xi$, the system is effectively at the percolation threshold p_c. The mass of the spanning cluster in this region is therefore given by (5.49), so that, $m \propto \ell^D$, where $\ell = \xi$, and therefore $m \propto \xi^D$. Consequently, the total mass of the spanning cluster is the mass of one such region multiplied with the number of regions:

$$M(p, L) \propto (\xi^D)(L/\xi)^d \propto \xi^{D-d} L^d \ . \tag{5.50}$$

Scaling Behavior of the Mass Above p_c We have therefore derived the behavior of the mass, $M(p, L)$, of the spanning cluster for $p > p_c$ for system sizes L that are much smaller or much larger than ξ:

$$M(p, L) \propto \begin{cases} L^D & \text{when } L \ll \xi \\ \xi^{D-d} L^d & \text{when } L \gg \xi \end{cases} . \tag{5.51}$$

This behavior can be rewritten in what we call the standard scaling form, with a scaling behavior and a cut-off function:

$$M(p, L) = L^D Y\left(\frac{L}{\xi}\right) , \tag{5.52}$$

where the cut-off function is:

$$Y(u) = \begin{cases} \text{constant} & u \ll 1 \\ u^{d-D} & u \gg 1 \end{cases} . \tag{5.53}$$

We will use this function form many times when we discuss the behavior of finite system sizes in the next chapter.

Exercises

Exercise 5.1 (Mass Scaling of Percolating Cluster)

(a) Find the mass $M(L)$ of the percolating cluster at $p = p_c$ as a function of L, for $L = 2^k, k = 4, \ldots, 11$.
(b) Plot $\log(M)$ as a function of $\log(L)$.
(c) Determine the exponent D.

Exercise 5.2 (Expressions for R_s^2) Show that

$$R_s^2 = \frac{1}{s}\langle \sum_i (\mathbf{r}_i - \mathbf{r}_{cm})^2 \rangle = \frac{1}{2}\frac{1}{s^2} \sum_{ij} (\mathbf{r}_i - \mathbf{r}_j)^2 ,$$

where the sum over ij are over all pairs of sites in a cluster of size s.

Hint Show that both expressions are equal to $\sum_i \mathbf{r}_i \cdot \mathbf{r}_i - \mathbf{r}_{cm} \cdot \mathbf{r}_{cm}$.

Exercise 5.3 (Correlation Function)

(a) Write a program to find the correlation function, $g(r, p, L)$ for $L = 256$.
(b) Plot $g(r, p, L)$ for $p = 0.55$ to $p = 0.65$ for $L = 256$.
(c) Find the correlation length $\xi(p, L)$ for $L = 256$ for the p-values used above.
(d) Plot ξ as a function of $p - p_c$, and determine ν.

Finite Size Scaling

6

In this chapter we will introduce the theory of finite size scaling and demonstrate how we can apply the theory to improve our measurements of the properties of percolation clusters. Usually, we attempt to measure properties of percolation system in the largest possible system we can simulate. Here, we demonstrate that if the system behaves according to simple scaling relations, it is instead much better to systematically vary the system size and the interpolate to infinite system sizes. This approach is generally called finite size scaling and we provide a thorough introduction to the theory and its applications to understand the scaling of the density of the spanning cluster, $P(p, L)$, the average cluster size, $S(p, L)$, and the percolation probability $\Pi(p, L)$.

How can we utilize a disadvantage, such as a finite system size, to an advantage? Usually, we have found a finite system size to be a hassle in simulations. We would like to find the general behavior, but we are limited by the largest finite system size we can afford to simulate. It may be tempting to put all our resources into one attempt—to make one simulation in a really large system. However, this is usually not a good strategy. Because we will then know that our results are limited by the system size, but we do not know to what degree the finite system size affects our result.

Instead, we will follow a different strategy: the strategy of finite size scaling. We will systematically increase the system size, measure the quantities we are interested in, and then try to extrapolate to an infinite system size. This has several advantages: It allows us to understand and estimate the errors in our predictions, and it allows us to use simulations of smaller systems. Indeed, it turns out that it is more important to do simulations in smaller systems, than only to try to simulate that largest system possible. However, for this to be effective, we need to have a theoretical understanding of finite size scaling [7].

The methods we develop here are powerful and can be generalized to many other experimental and computational situations. In many experiments it is also tempting to try to perform the perfect experiment by reducing noise or measurement

A. Malthe-Sørenssen, *Percolation Theory Using Python*, Lecture Notes
in Physics 1029, https://doi.org/10.1007/978-3-031-59900-2_6

errors. For example, we may perform an experiment where we need to make the experimental system as horizontal as possible, because deviations from a horizontal system would introduce errors. Instead of trying to make the system as horizontal as possible, we may instead systematically vary the orientation, and then extrapolate to the case when the system is perfectly horizontal. This allows us to control the uncertainty. Of course, we cannot vary all possible uncertainties in an experiment or a simulation, but this alternative mindset provides us with a new tool in our toolbox, and a new way to deal with uncertainties.

In practical situations, we will always be limited by finite system sizes. If you measure the size of earthquakes in the Earth's crust, your results are limited by the thickness of the crust or by the extent of a homogeneous region. If you simulate a molecular system, you are definitely limited by the number of atoms you can include in your simulation. Thus, better insight into how we can systematically vary the system size and use this to gain insight are general tools of great utility.

Here, you will learn how to systematically vary system size L in order to find good estimates for exponents and percolation thresholds. Indeed, my hope is that you will see that finite size scaling is a powerful tool that can be used both theoretically and computationally. To introduce this tool, we need to address specific examples that can help build our intuition and shape our mindset. We will therefor start from a few examples, such as the finite size scaling for the density of the spanning cluster, $P(p, L)$, and then apply the method to a new case, the percolation probability $\Pi(p, L)$.

6.1   General Aspects of Finite Size Scaling

We have found that a percolation system is described by three length-scales: the size of a site, the system size L, and the correlation length ξ. Finite size scaling addresses the change in behavior of a system as we change the system size L. Typically, we divide the behavior into two categories:

- When the system size L is much smaller than the correlation length ξ, $L \ll \xi$, the system appears to be on the percolation threshold.
- When L is much larger than ξ, $L \gg \xi$, the geometry is essentially homogeneous at lengths longer than ξ.

We will then address the behavior close to p_c. In the case of percolation, we usually assume that the behavior is a power-law in $p - p_c$. For example, the mass $M(p; L)$ of the spanning cluster:

$$M(p) \propto (p - p_c)^{-x} , \tag{6.1}$$

where the exponent x determines the behavior close to p_c.

The general approach to finite size scaling is to make a scaling ansatz, that is, an assumption about how the system behaves, which typically consists of a scaling term and a cut-off function, as you have seen several times in this book:

$$M(p, L) = L^{\frac{x}{\nu}} f\left(\frac{L}{\xi}\right) , \tag{6.2}$$

where $f(u)$ is an unknown function. (Sometimes we instead make the assumption $M(p, L) = \xi^{x/\nu} \tilde{f}(L/\xi)$. We leave it to the reader to demonstrate that these assumptions are equivalent.)

We will then apply our insight into the particulars of the system to infer the behavior in the limits when $\xi \gg L$, and $\xi \ll L$ to determine the form of the scaling function $f(u)$, and use this functional form as a tool to study the behavior of the system. We will explain this reasoning through three examples: The case of $P(p, L)$, the case of $S(p, L)$ and the case of $\Pi(p, L)$.

6.2   Finite Size Scaling of $P(p, L)$

Measuring $P(p, L)$ **for finite** L Let   us now apply this methodology to study the behavior of the density of the spanning cluster, $P(p, L)$, for finite system sizes. First, we generate a plot of $P(p, L)$ for various values of L using the following program:

```
import numpy as np
import matplotlib.pyplot as plt
from scipy.ndimage import measurements
LL = [25,50,100,200]
p = np.linspace(0.4,0.75,50)
nL = len(LL)
nx = len(p)
Ni = np.zeros(nx)
P = np.zeros((nx,nL),float)
for iL in range(nL):
    L = LL[iL]
    N = int(2000*25/L)
    for i in range(N):
        z = np.random.rand(L,L)
        for ip in range(nx):
            m = z<p[ip]
            lw, num = measurements.label(m)
            perc_x = np.intersect1d(lw[0,:],lw[-1,:])
            perc = perc_x[np.where(perc_x>0)]
            if (len(perc)>0):
                Ni[ip] = Ni[ip] + 1
                area = measurements.sum(m, lw, perc[0])
                P[ip,iL] = P[ip,iL] + area
    P[:,iL] = P[:,iL]/((L*L)*N)
for iL in range(nL):
    L = LL[iL]
    plt.plot(p,P[:,iL],label="L = "+str(L))
```

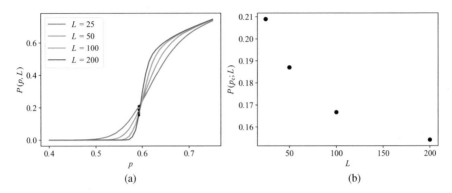

Fig. 6.1 (a) Plot of $P(p, L)$. (b) Plot of $P(p_c; L)$ as a function of L

```
plt.ylabel('$P(p,L)$')
plt.xlabel('$p$')
plt.legend()
```

The resulting plot of $P(p, L)$ is shown in Fig. 6.1. We see that as L increases, $P(p, L)$ approaches the shape expected in the limit when $L \to \infty$. We can see how it approaches this limit by finding the value of $P(p_c, L)$ as a function of L. We expect this value to go to zero as L increases. Figure 6.1b shows how $P(p_c, L)$ approaches zero. Let us see if we can develop a theoretical prediction for this behavior and check if our measured results confirm the prediction.

Finite Size Effects in $P(p, L)$ We know that $P(p) \propto (p - p_c)^\beta$ and $\xi \propto |p - p_c|^{-\nu}$, so that

$$P(p) \propto (p - p_c)^\beta \propto \xi^{-\beta/\nu} . \tag{6.3}$$

This is valid in the limit when $L \to \infty$, that is, when $L \gg \xi$. In the limit when $L \ll \xi$, which eventually will occur as p approaches p_c and ξ diverges, we see from Fig. 6.1 that $P(p_c, L)$ depends on L. In this case, we have previously found that

$$P(p, L) \simeq P(p_c, L) = \frac{M(p_c, L)}{L^2} \propto \frac{L^D}{L^d} \propto L^{D-d} \propto L^{-\beta/\nu} . \tag{6.4}$$

Combined, we therefore have the behavior

$$P(p, L) \propto \begin{cases} \xi^{-\beta/\nu} & \text{when } L \gg \xi \\ L^{-\beta/\nu} & \text{when } L \ll \xi \end{cases} . \tag{6.5}$$

Finite Size Scaling Ansatz The fundamental idea of finite size scaling is then to *assume* a particular form of a function that encompasses this behavior both when $\xi \ll L$ and $\xi \gg L$, by rewriting the expression for $P(p, L)$ as

$$P(p, L) = L^{-\beta/\nu} f(L/\xi) . \tag{6.6}$$

Where we have assumed that the only relevant length scales are L and ξ, and that the function therefore only can depend on a ratio between these two length scales. How must the function $f(u)$ behave for this general form to reduce to Eqs. (6.3) and (6.4)?

First, we see that when $\xi \gg L$ the function $f(L/\xi)$ should be a constant, that is, $f(u)$ is a constant when $u \ll 1$. Second, we see that when $\xi \ll L$, we need the function $f(L/\xi)$ to cancel all the L-dependency in order to find the relation in Eq. (6.3):

$$P(p, L) = L^{-\beta/\nu} f(L/\xi) = \xi^{-\beta/\nu} . \tag{6.7}$$

This will occur if and only if $f(u)$ is a power-law, that is, $f(u) = u^a$. In order to cancel the L-dependency, the power-law exponent for the L-term must be zero:

$$P(p, L) \propto L^{-\beta/\nu}(L/\xi)^a \propto L^{-\beta/\nu+a}\xi^{-a} \propto \xi^{-\beta/\nu} \tag{6.8}$$

$$\Rightarrow -\beta/\nu + a = 0 \Rightarrow a = \beta/\nu . \tag{6.9}$$

Indeed, we could have used this in order to *find* the exponent in the relation $\xi^{-\beta/\nu}$. It would simply have been enough to assume that $P(p, L) \propto \xi^x$ for some exponent x in the limit of $\xi \ll L$.

In order to satisfy these conditions, the scaling form of $P(p, L)$ must therefore be

$$P(p, L) = L^{-\beta/\nu} f(L/\xi) , \tag{6.10}$$

where

$$f(u) = \begin{cases} \text{const.} & \text{for } u \ll 1 \\ u^{\beta/\nu} & \text{for } u \gg 1 \end{cases} \tag{6.11}$$

Testing the Scaling Ansatz We can now test the scaling ansatz by plotting $P(p, L)$ according to the ansatz, following a strategy similar to what we developed for $n(s, p)$. We rewrite the scaling function $P(p, L) = L^{-\beta/\nu} f(L/\xi)$ in terms of $|p - p_c|$ by inserting $\xi = \xi_0 |p - p_c|^{-\nu}$:

$$P(p, L) = L^{-\beta/\nu} f(L/\xi) \tag{6.12}$$

$$= L^{-\beta/\nu} f(L\xi_0 |p - p_c|^\nu) \tag{6.13}$$

Fig. 6.2 Scaling data
collapse plot of $P(p, L)$ with
$L^{1/\nu}(p - p_c)$ along the
x-axis and $L^{\beta/\nu} P(p, L)$
along the y-axis

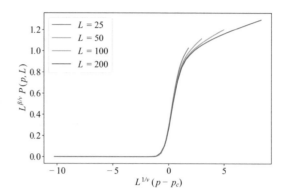

$$= L^{-\beta/\nu} f((\xi_0 L^{1/\nu}(p - p_c))^\nu) \tag{6.14}$$

$$= L^{-\beta/\nu} \tilde{f}(L^{1/\nu}(p - p_c)) . \tag{6.15}$$

Which again can be rewritten as

$$L^{\beta/\nu} P(p, L) = \tilde{f}(L^{1/\nu}(p - p_c)) . \tag{6.16}$$

Consequently, if we plot $L^{1/\nu}(p - p_c)$ along the x-axis and $L^{\beta/\nu} P(p, L)$ along the y-axis, we expect the data from simulations for various L-values to fall onto a common curve, the curve $f(u)$. This is illustrated in Fig. 6.2, which shows that the measured data is consistent with the scaling ansatz. We call such as plot a *scaling data collapse* plot.

Comparing to Theory at $p = p_c$ Finally, we can now use this theory to understand the behavior for $P(p_c, L)$. In this case we find that $P(p_c, L) = cL^{-\beta/\nu}$. We can therefore measure $-\beta/\nu$ from the plot of $P(p_c, L)$ in Fig. 6.1. While the data in this figure is too poor to produce a reliable result, the figure demonstrates the principle.

Varying L to Gain Insight   The take-home message is that instead of trying to simulate one single simulation with as large L as possible, we instead vary L systematically and then use this variation to estimate the relevant exponents ν and β. The methods demonstrated here usually provide much better results in term of precision of the exponents than a direct measurement for a large system size.

Alternative Approaches   We could instead have started with a scaling ansatz of $P(p, L) = (p - p_c)^\beta g(L/\xi) = \xi^{-\beta/\nu} g(L/\xi)$. However, the end result would be the same. We leave this as an exercise for the eager reader.

6.3   Average Cluster Size

We can characterize the distribution of cluster sizes using *moments of the cluster number distribution*. The k-th moment $M_k(p, L)$ is defined as:

$$M_k(p, L) = \sum_{s=1}^{\infty} s^k n(s, p; L) \, . \tag{6.17}$$

We have already introduced the second moment, $M_2(p, L)$, which we called the average cluster size, $S(p, L)$.

$$S(p, L) = M_2(p, L) = \sum_{s=1}^{\infty} s^2 n(s, p; L) \, . \tag{6.18}$$

Now, let us see if we can apply the finite-size scaling approach to develop a scaling theory for $S(p, L)$. First, we will measure $S(p, L)$, and then develop and test a scaling theory.

Measuring Moments of the Cluster Number Density

How would we measure $S(p, L)$? We recall that we measure the cluster number density from

$$\overline{n(s, p; L)} = \frac{N_s}{L^d} \, , \tag{6.19}$$

where N_s is the number of clusters of size s. Thus we can estimate $S(p, L)$ from:

$$\overline{S(p, L)} = \sum_{s=1}^{\infty} s^2 \overline{n(s, p; L)} = \sum_{s=1}^{\infty} s^2 \frac{N_s}{L^d} \, . \tag{6.20}$$

We realize that we can perform this sum by summing over all possible s and then including how many clusters we have for a given s, or we can alternatively sum over all the observed clusters s_i. (Try to convince yourself that this is the same by looking at a sequence of clusters of sizes 1, 2, 1, 5, 1, 2.). Thus, we can estimate the second moment from the sum:

$$\overline{S(p, L)} = \sum_{i} s_i^2 / L^2 \, . \tag{6.21}$$

And similarly by summing over s_i^k for the k-th moment. We implement this in the following program:

```python
import numpy as np
import matplotlib.pyplot as plt
from scipy.ndimage import measurements
LL = [25,50,100,200]
p = np.linspace(0.4,0.75,50)
nL = len(LL)
nx = len(p)
S = np.zeros((nx,nL),float)
for iL in range(nL):
    L = LL[iL]
    M = int(2000*25/L)
    for i in range(M):
        z = np.random.rand(L,L)
        for ip in range(nx):
            m = z<p[ip]
            lw, num = measurements.label(m)
            labelList = np.arange(lw.max() + 1)
            area = measurements.sum(m, lw, labelList)
            # Remove spanning cluster by setting area to zero
            perc_x = np.intersect1d(lw[0,:],lw[-1,:])
            perc = perc_x[np.where(perc_x>0)]
            if (len(perc)>0):
                area[perc[0]] = 0
            S[ip,iL] = S[ip,iL] + np.sum(area*area)
    S[:,iL] = S[:,iL]/(L**2*M)
# Plotting the results
plt.figure(figsize=(6,4))
for iL in range(nL):
    L = LL[iL]
    lab = "$L="+str(L)+"$"
    plt.plot(p,S[:,iL],label=lab)
plt.ylabel('$S(p,L)$')
plt.xlabel('$p$')
plt.legend()
```

The resulting plot of $S(p, L)$ as a function of p for various values of L is shown in Fig. 6.3.

Scaling Theory for $S(p, L)$

How can we understand these plots and how can we develop a theory for $S(p, L)$? We previously found that S diverges as p approaches p_c:

$$S(p) = S_0|p - p_c|^{-\gamma} , \tag{6.22}$$

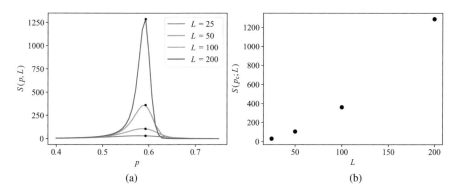

Fig. 6.3 (a) Plot of $S(p, L)$. (b) Plot of $S(p_c; L)$ as a function of L

where the exponent is $\gamma = 43/18$ for $d = 2$. Following the approach for finite-size scaling introduced above, we introduce the finite size L through a scaling function $f(L/\xi)$, giving us a finite-size scaling ansatz (our hypothesis):

$$S(p, L) = S_0|p - p_c|^{-\gamma} f\left(\frac{L}{\xi}\right). \tag{6.23}$$

We rewrite the first expression by introducing $\xi = \xi_0|p - p_c|^{-\nu}$ so that $S_0|p - p_c|^{-\gamma} = \xi^{\gamma/\nu}$, giving:

$$S(p, L) = \xi^{\gamma/\nu} f\left(\frac{L}{\xi}\right). \tag{6.24}$$

Now, we see from Fig. 6.3 that when $p = p_c$, $S(p_c, L)$ does not diverge, but is limited by L, as we would expect for a finite system. Thus we know that in the limit when $p \to p_c$, $S(p, L)$ can only depend on L. This implies that the function $f(L/\xi)$ in this limit must be so that the ξ in $f(L/\xi)$ cancels the $\xi^{\gamma/\nu}$ in front of it. This can only happen if $f(L/\xi) \propto (L/\xi)^{\gamma/\nu}$:

$$S(p, L) \propto \xi^{\gamma/\nu} \left(\frac{L}{\xi}\right)^{\gamma/\nu} \propto L^{\gamma/\nu}. \tag{6.25}$$

Thus, we have found that $S(p_c, L) \propto L^{\gamma/\nu}$.

This allows us to write the scaling form of $S(p, L)$ in a different way:

$$S(p, L) = L^{\gamma/\nu} g\left(\frac{L}{\xi}\right). \tag{6.26}$$

Fig. 6.4 A data-collapse plot
of the rescaled average cluster
size $L^{-\gamma/\nu} S(p, L)$ as a
function of $L^{1/\nu}(p - p_c)$ for
various L

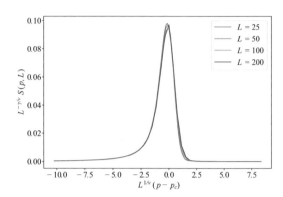

We can test this prediction by plotting $S(p, L)L^{-\gamma/\nu}$ as a function of L/ξ:

$$S(p, L)L^{-\gamma/\nu} = g\left(\frac{L}{\xi}\right) = g\left(L(p - p_c)^{-\nu}\right) \tag{6.27}$$

$$= g\left(\left(L^{1/\nu}(p - p_c)\right)^{\nu}\right) = \tilde{g}\left(L^{1/\nu}(p - p_c)\right). \tag{6.28}$$

The resulting plot is shown in Fig. 6.4, which indeed demonstrates that the measured
data is consistent with the scaling theory. Success!

6.4   Percolation Threshold

Finally, we will demonstrate one of the most elegant applications of finite-size
scaling theory to the percolation probability $\Pi(p, L)$ and to see how a finite system
size will affect the effective percolation threshold.

Measuring the Percolation Probability $\Pi(p, L)$

We can measure the percolation probability for a set of finite system sizes using the
methods we developed previously. Here, we have implemented the measurement in
the following program which is very similar to the program developed to measure
$P(p, L)$

```
import numpy as np
import matplotlib.pyplot as plt
from scipy.ndimage import measurements
LL = [25,50,100,200]
p = np.linspace(0.4,0.75,50)
nL = len(LL)
nx = len(p)
```

```
Ni = np.zeros((nx,nL),float)
Pi = np.zeros((nx,nL),float)
for iL in range(nL):
    L = LL[iL]
    N = int(2000*25/L)
    for i in range(N):
        z = np.random.rand(L,L)
        for ip in range(nx):
            m = z<p[ip]
            lw, num = measurements.label(m)
            perc_x = np.intersect1d(lw[0,:],lw[-1,:])
            perc = perc_x[np.where(perc_x>0)]
            if (len(perc)>0):
                Ni[ip,iL] = Ni[ip,iL] + 1
    Pi[:,iL] = Ni[:,iL]/N
for iL in range(nL):
    L = LL[iL]
    lab = "$L="+str(L)+"$"
    plt.plot(p,Pi[:,iL],label=lab)
plt.ylabel('$\Pi(p,L)$')
plt.xlabel('$p$')
plt.legend()
```

The resulting plot of $\Pi(p, L)$ for various values of L is shown in Fig. 6.5.

Measuring the Percolation Threshold p_c

Let us now assume that we do not a priori know p_c or any of the scaling exponents. How can we use this data-set to estimate the value for p_c?

The simplest approach may be to estimate p_c as the value for p that makes $\Pi(p, L) = 1/2$. This corresponds to intersection between the horizontal line

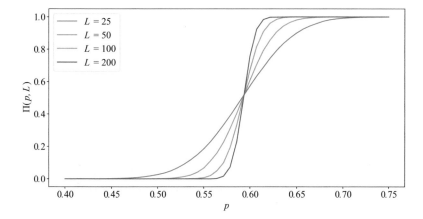

Fig. 6.5 Plot of $\Pi(p, L)$

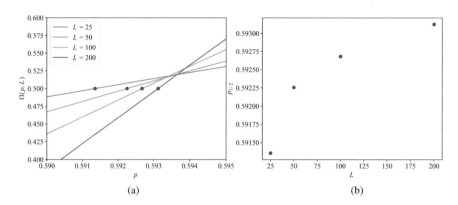

Fig. 6.6 (**a**) Plot of $\Pi(p, L)$. (**b**) Plot of $p_{1/2}$ as a function of L

$\Pi = 1/2$ and the curves in Fig. 6.5. This is illustrated in Fig. 6.6. Here, we
have also plotted $p_{1/2}$ as a function of L, where $p_{1/2}$ is the value for p so that
$\Pi(p_{1/2}, L) = 1/2$. These values for $p_{1/2}$ are calculated by a simple interpolation
as illustrated in the following program. (Notice that as usual in this book, we do not
aim for high precision in this program. The simulations are for small system sizes
and few samples, but are meant to illustrate the principle and be reproduceable for
you.)

```
for iL in range(nL):
    ipc = np.argmax(Pi[:,iL]>0.5) # Find first i where Pi>0.5
    # Interpolate from ipc-1 to ipc to find intersection
    ppc = p[ipc-1] + (0.5-Pi[ipc-1,iL])*\
        (p[ipc]-p[ipc-1])/(Pi[ipc,iL]-Pi[ipc-1,iL])
    Pic = 0.5
    plt.plot(LL[iL],ppc,'o')
plt.xlabel('$L$')
plt.ylabel('$p_{1/2}$')
```

From Fig. 6.6 we see that as L increases the value for $p_{1/2}$ gradually approaches
p_c. Well, we cannot really see that it is approaching p_c, but we guess that it will.
However, in order extrapolate the curve to infinite L we need to develop a theory
for how $p_{1/2}$ behaves. We need to develop a finite size scaling theory for $\Pi(p, L)$.

Finite-Size Scaling Theory for $\Pi(p, L)$

We apply the same method as before to develop a theory for $\Pi(p, L)$. First. we
notice that at p_c $\Pi(p_c, L)$ does not either diverge or go to zero. This means that

$\Pi(p, L)$ cannot be a function of ξ alone, but instead must have the scaling form:

$$\Pi(p, L) = \xi^0 f\left(\frac{L}{\xi}\right) . \tag{6.29}$$

We rewrite this in terms of $(p - p_c)$ by inserting $\xi = \xi_0 |p - p_c|^{-\nu}$:

$$\Pi(p, L) = f\left(L\xi_0|p - p_c|^{\nu}\right) = f\left(\xi_0\left(L^{1/\nu}(p - p_c)\right)^{\nu}\right) . \tag{6.30}$$

We introduce a new function $\Phi(u) = f\left(\xi_0 u^{1/\nu}\right)$:

$$\Pi(p, L) = \Phi\left(L^{1/\nu}(p - p_c)\right) . \tag{6.31}$$

This is our finite-size scaling ansatz (theory).

Estimating p_c Using the Scaling Ansatz

How can we use this theory to estimate p_c? We follow a technique similar to what we used above: We find the value p_x that makes $\Pi(p_x, L) = x$. Above, we did this for $x = 1/2$, but we can do this more generally. Actually, as $L \rightarrow \infty$, we expect any such p_x to converge to p_c. We notice from above that p_x is a function of L: $p_x = p_x(L)$.

We insert this into the scaling ansatz:

$$x = \Phi\left((p_x(L) - p_c) L^{1/\nu}\right) , \tag{6.32}$$

which can be solved as

$$(p_x - p_c)L^{1/\nu} = \Phi^{-1}(x) = C_x , \tag{6.33}$$

where it is important to realize that the right hand side, C_x, is a number which only depends on x and not on L. We can therefore rewrite this as

$$p_x - p_c = C_x L^{-1/\nu} . \tag{6.34}$$

If we know ν, we see that this gives a method to estimate the value of p_c. Figure 6.7 shows a plot of $p_{1/2} - p_c$ as a function of $L^{-1/\nu}$ for $\nu = 4/3$. We can use this plot to extrapolate to find p_c in the limit when $L \rightarrow \infty$ as indicated in the plot. The resulting value for p_c extrapolated from $L = 50, 100, 200$ is $p_c = 0.5935$, which is surprisingly good given the small system sizes and small sample sizes used for this estimate. (The best known value is $p_c = 0.5927$). This demonstrates the power of finite size scaling.

Fig. 6.7 Plot of $p_{1/2}$ as a function of $L^{-1/\nu}$. The dashed line indicates a linear fit to the data for $L = 50, 100, 200$. The extrapolated value for p_c at $L \to \infty$ is $p_c = 0.5935$

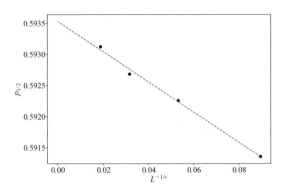

Estimating p_c and ν Using the Scaling Ansatz

However, this approach depends on us knowing the value for ν. What if we did not know neither ν nor p_c? How can we estimate both from the scaling ansatz? One alternative is to generate plots of p_x as a function of $L^{-1/\nu}$ for several values of x. Then we adjust the values of ν until we get a straight line, in that case we can read of the intersect with the p_x axis as the value for p_c.

However, we can do even better by noticing a trick: For two x values x_1 and x_2, we get

$$dp = p_{\Pi=x_1}(L) - p_{\Pi=x_2}(L) = (C_{x_1} - C_{x_2})L^{-\nu} , \qquad (6.35)$$

and we can therefore plot $\log(dp)$ as a function of $\log(L)$ to get ν, and then use this to estimate p_c. As an exercise, the reader is encouraged to demonstrate that this scaling ansatz is valid for $d = 1$, and in this case find C_x explicitly.

Exercises

Exercise 6.1 (Finite-Size Scaling in One Dimension)

(a) Show that the scaling ansatz for $\Pi(p, L)$ is valid for $d = 1$.
(b) Find an explicit expression for C_x for $d = 1$.

Exercise 6.2 (Finite-Size Scaling in Two Dimensions)   In this exercise we will use the scaling ansatz to provide estimates of ν, p_c and the average percolation probability $\langle p \rangle$ in a system of size L.

We define p_x so that $\Pi(p_x, L) = x$. Notice that p_x is a function of system size L used for the simulation.

(a) Find p_x for $x = 0.3$ and $x = 0.8$ for $L = 25, 50, 100, 200, 400, 800$. Plot p_x as a function of L.

According to the scaling theory we have

$$p_{x_1} - p_{x_2} = \left(C_{x_1} - C_{x_2}\right) L^{-1/\nu} . \qquad (6.36)$$

(b) Plot $\log(p_{0.8} - p_{0.3})$ as a function of $\log(L)$ to estimate the exponent ν. How does it compare to the exact result?

In the following, please use the exact value $\nu = 4/3$. The scaling theory also predicted that

$$p_x = p_c + C_x L^{-1/\nu} . \qquad (6.37)$$

(c) Plot p_x as a function of $L^{-1/\nu}$ to estimate p_c. Generate a data-collapse plot for $\Pi(p, L)$ to find the function $\Phi(u)$ described above.
(d) Plot $\Pi'(p, L)$ as a function of p for the various L values used above. Generate a data-collapse plot of $\Pi'(p, L)$. Find $\langle p \rangle$ and plot $\langle p \rangle$ as a function of $L^{-1/\nu}$ to find p_c.

Exercise 6.3 (Finite Size Scaling of $n(s, p_c, L)$)

(a) Develop a finite size scaling ansatz/theory for $n(s, p_c, L)$. You should provide arguments for the behavior in the various limits.
(b) Plot $n(s, p_c, L)$ as a function of s for $L = 100, 200, 400, 800$.
(c) Demonstrate the validity of the scaling theory by producing a data-collapse plot for $n(s, p_c, L)$.

Renormalization

7

In this chapter we will introduce the powerful theoretical methods of renormalization. The fundamental idea is that at $p = p_c$, a rescaling of the system does not change the most important features. By a rescaling we typically mean a coarse-graining of the system, such as merging 2×2 cells into a single cell. The rule we use to choose the occupation probability of the new, coarse-grained cell, p', is a function of the probability p of the original lattice, $p' = R(p)$. In renormalization theory, we use properties of this mapping, $R(p)$, to deduce properties of the system such as critical exponents. In this chapter, you will be introduced to the fundamentals of renormalization theory in the context of percolation systems, in which the geometric nature of the remapping allow us to build intuition about renormalization as a concept. We will also apply the theory to different lattice structures and for one, two and three-dimensional systems.

We have now learned that when p approaches p_c, the correlation length grows to infinity, and the spanning cluster becomes a self-similar fractal structure. This implies that the spanning cluster at p_c has statistical self-similarity: if we cut out a piece of the spanning cluster, and rescale the lengths in the system, the rescaled system will have the same statistical geometrical properties as the original system. In particular, the rescaled system will have the same mass scaling relation: it will also be a self-similar fractal with the same scaling properties.

What happens when $p \neq p_c$? In this case, there will be a finite correlation length, ξ, and a rescaling of the lengths in the system implies that the correlation length is also rescaled. A rescaling by a factor b corresponds to making a coarse-graining over b^d sites in order to form the new lattice. Now, we will simply assume that this also implies that the correlation length is reduced by a factor b: $\xi' = \xi/b$. After a few iterations of this rescaling procedure, the correlation length will correspond to the lattice size and the lattice will be uniform.

We could have made this argument even simpler by initially stating that we divide the system into parts that are larger than the correlation length. Again, this would lead to a system that is homogeneous from the smallest lattice spacing an upwards.

© The Author(s) 2024
A. Malthe-Sørenssen, *Percolation Theory Using Python*, Lecture Notes in Physics 1029, https://doi.org/10.1007/978-3-031-59900-2_7

We can conclude that when $p < p_c$, the system behaves as a uniform, unconnected system and when $p > p_c$, the system is uniform and connected.

The argument we have sketched above is the essence of *the renormalization group argument*. It is only exactly at $p = p_c$ that an iterative rescaling is a non-trivial fix point: the system iterates onto itself because it is a self-similar fractal. When p is away from p_c, rescaling iterations will make the system progressively more homogeneous, and effectively bring the rescaled p towards either 0 or 1.

In this chapter we will provide an introduction to the theoretical framework for renormalization. This is a powerful set of techniques, introduced for equilibrium critical phenomena by Kadanoff [19] in 1966 and by Wilson [39] in 1971. Wilson later received the Nobel prize for his work on critical phenomena.

7.1   The Renormalization Mapping

What happens when we coarse-grain a percolation system? What does it mean to coarse-grain? It means that we replace a 2×2 cell with a single cell using a specified rule, which aims at retaining connectivity. An example of such a rule is given in Fig. 7.1. For each possible 2×2 configuration, we show if it maps onto an occupied or an empty cell. Let us now apply this rule to a 64×64 system for three different values of p as illustrated in Fig. 7.2. We iterate the procedure several time, reducing the system size with a factor of 2 each time.

Behavior Through Iterations What happens in this system? When $p = p_c$, then ξ is infinte. This means that ξ does not change we divide the system size by 2. We see that the system appears similar throughout the iteration, and the final single site is occupied. This is because the system is a self-similar fractal and does not change significantly through the iterations. What happens when $p > p_c$? In this case, we see that the system becomes more homogeneous through each iteration and eventually the whole system is filled. This means that the effective percolation probability becomes higher through the iterations. Similarly, when $p > p_c$, the system becomes more homogeneous, but also more empty, as the iterations proceed.

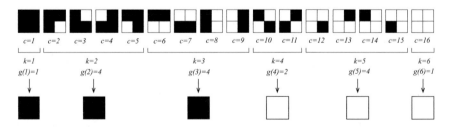

Fig. 7.1 Illustration of a renormalization rule for a site percolation problem with a rescaling $b = 2$. The top row shows the 16 configurations c. The middle row show the 6 classes k with multiplicities $g(k)$

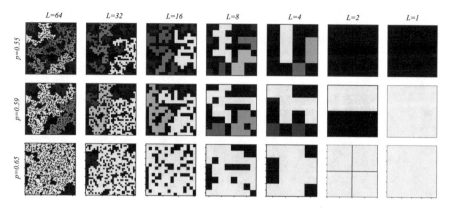

Fig. 7.2 Illustration of averaging using a rescaling $b = 2$, so that a cell of size $b \times b = 2 \times 2$ is reduced to a single site, producing a "renormalized" system of size $L/2$. We iterate the procedure until $L = 1$, that is, only a single site is left

This means that the effective percolation probability becomes lower through the iterations.

Renormalization Means Changing the Occupation Probability   In the original lattice the occupation probability is p. However, through our coarse-graining procedure, we may change the occupation probability for the new, averaged sites. We will therefore call the new occupation probability p', the probability to occupy a renormalized site. We write the mapping between the original and the new occupation probabilities as

$$p' = R(p) , \qquad (7.1)$$

where the renormalization function $R(p)$, which provides the mapping, depends on the details of the rule used for renormalization.

Selecting a Renormalization Rule   There are many choices for the mapping between the original and the renormalized lattice. We have illustrated a particular mapping with a rescaling $b = 2$ in Fig. 7.1. Such a mapping describes how each of the $4^2 = 16$ possible configurations c of the 2×2 system is mapped onto a 1×1 single site through a function $f(c)$, where $f(c)$ is 1 if the new site is occupied and 0 if it is empty. The renormalization mapping is then

$$R(p) = \sum_c P(c) f(c) , \qquad (7.2)$$

where $P(c)$ is the probability for configuration c. It is often practical to organize the configurations into classes k, where each class has the same number of occupied sites and hence the same probability $P(k)$, and the number of configurations in class

k is called the multiplicity $g(k)$ of the class. Expressed in terms of the classes k, the renormalization mapping is

$$R(p) = \sum_k g(k) P(k) f(k) \ . \tag{7.3}$$

For the particular mapping provided in Fig. 7.1, the renormalization mapping becomes

$$\begin{aligned} R(p) &= 1 \cdot p^4 \cdot 1 + 4 \cdot p^3 (1-p)^1 \cdot 1 + 4 \cdot p^2 (1-p)^2 \cdot 1 \\ &\quad + 2 \cdot p^2 (1-p)^2 \cdot 0 + 4 \cdot p^1 (1-p)^3 \cdot 0 + 1 \cdot (1-p)^4 \cdot 0 \tag{7.4} \\ &= p^4 + 4p^3 (1-p) + 4p^2 (1-p^2) \ . \end{aligned}$$

This illustrates a particular rule, but there are many possible rules. Usually, we want to ensure that important aspects of the percolation system is preserved by the mapping. For example, we would want the mapping to conserve connectivity. That is, we would like to ensure that

$$\Pi(p, L) = \Pi(p', \frac{L}{b}) \ . \tag{7.5}$$

However, even though we may ensure this on the level of the mapping, this does not ensure that the mapping actually conserves connectivity when applied to a large cluster. It may, for example, connect clusters that were unconnected in the original lattice, or disconnect clusters that were connected, as illustrated in Fig. 7.3.

Properties of the Renormalization Mapping First, we will not consider the details of the renormalization mapping $p' = R(p)$, but instead assume that such

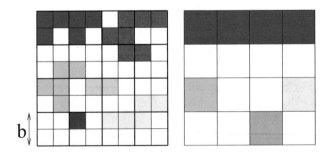

Fig. 7.3 Illustration of a single step of renormalization on an 8×8 lattice of sites. We see that the renormalization procedure introduces new connections: the blue cluster is now much larger than in the original. However, the procedure also removes previously existing connections: the original yellow cluster is split into two separate clusters

a map exists and study its qualitative features. Then we will address detailed properties of the renormalization mapping through two worked examples.

For any choice of mapping, the rescaling will result in a change in the correlation length ξ:

$$\xi' = \xi(p') = \frac{1}{b}\xi(p) \ . \tag{7.6}$$

We will use this relation to address the behavior of *fixpoints* of the mapping.

Fixpoint A *fixpoint* of a mapping $R(p)$ is a point p^* that does not change when the mapping is applied. That is

$$p^* = R(p^*) \ . \tag{7.7}$$

Trivial Fixpoints At a fixpoint, the iteration relation for the correlation length becomes:

$$\xi(p^*) = \frac{\xi(p^*)}{b} \ . \tag{7.8}$$

The only possible solutions for this equation are that $\xi = 0$ or $\xi = \infty$. We call the case when $\xi = 0$ a *trivial fixed* point. There are two trivial fixed points for any renormalization mapping at $p = 0$ and at $p = 1$.

Stable and Unstable Fixpoints Let us assume that there exists a *non-trivial fixpoint* p^*, and let us address the behavior for p close to p^*. We notice that for any finite ξ, iterations by the renormalization relation will reduce ξ. That is, both for $p < p^*$ and for $p > p^*$ iterations will make ξ smaller. This implies that iterations will take the system further away from the non-trivial fixpoint, where the correlation length is infinite. The non-trivial fixpoint is therefore an *unstable fixpoint*. Similarly, for p close to a trivial fixpoint, where $\xi = 0$, iterations will decrease ξ, and the renormalized system will move closer to the fixpoint in each iteration. The trivial fixpoint is therefore *stable*.

Graphical Interaction of the Renormalization Relation Iterations by the renormalization relation $p' = R(p)$ may be studied on the graph $R(p)$, as illustrated in Fig. 7.4. Consecutive iterations take the system along the arrows illustrated in the figure. Notice that the line $p' = p$ is drawn as a dotted reference line. In the figure, the two end points, $p = 0$ and $p = 1$ are the only stable fixpoints, and the point p^* is the only unstable fixpoint. The actual shape of the function $R(p)$ depends on the renormalization rule, and the shape may be more complex than what is illustrated in Fig. 7.4.

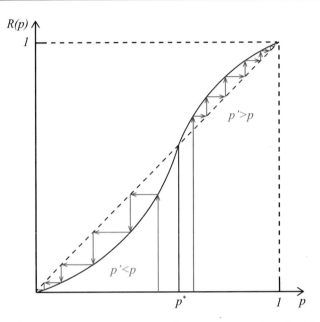

Fig. 7.4 Illustration the renormalization mapping $p' = R(p)$ as a function of p. The non-trivial fixpoint $p^* = R(p^*)$ is illustrated. Two iterations sequences are illustrated by the lines with arrows. Let us look at the path starting from $p > p^*$. Through the first application of the mapping, we read off the resulting value of p'. This value will then be the input value for the next application of the renormalization mapping. A fast way to find the corresponding position along the p axis is to reflect the p' value from the line $p' = p$ shown as a dotted line. This gives the new p value, and the mapping is applied again producing yet another p' which is even further from p^*. With the drawn shape of $R(p)$ there is only one non-trivial fixpoint, which is unstable

Iterating the Renormalization Mapping

We are now ready for a more quantitative argument for the effect of iterations through the renormalization mapping $R(p)$. First, we notice that the non-trivial fixpoint corresponds to the percolation threshold of the renormalization model, since the correlation length is diverging for this value of p. (This does not imply that p^* is equal to p_c. As we shall see, p^* depends on the choice of $R(p)$).

We will now assume that $R(p)$ is differentiable, which it should be since $R(p)$ is based on sums of polynomials of p and $1 - p$. Let us study the behavior close to p^* through a Taylor expansion of the mapping $p' = R(p)$. First, we notice that

$$p' - p^* = R(p) - R(p^*) , \qquad (7.9)$$

because $p' = R(p)$ and $p^* = R(p^*)$. The Taylor expansion of $R(p)$ for a p close to p^* is:

$$R(p) = R(p^*) + R'(p^*)(p - p^*) + \mathcal{O}(p - p^*)^2 . \qquad (7.10)$$

If we define $\Lambda = R'(p^*)$, we get that to first order in $p - p^*$:

$$p' - p^* \simeq \Lambda(p - p^*) \,, \tag{7.11}$$

We see that the value of Λ characterizes the fixpoint. For $\Lambda > 1$ the new point p' will be further away from p^* than the initial point p. Consequently, the fixpoint is unstable. By a similar argument, we see that for $\Lambda < 1$ the fixpoint is stable. For $\Lambda = 1$ we call the fixpoint a marginal fixpoint.

Let us now assume that the fixpoint is indeed the percolation threshold. In this case, when p is close to p_c, we know that the correlation length is

$$\xi(p) = \xi_0(p - p_c)^{-\nu} \,, \tag{7.12}$$

for the initial point, and

$$\xi(p') = \xi_0(p' - p_c)^{-\nu} \tag{7.13}$$

for the renormalized point. We will now use (7.11) for $p^* = p_c$, giving

$$p' - p_c = \Lambda(p - p_c) \,. \tag{7.14}$$

Inserting this into (7.13) gives

$$\xi(p') = \xi_0(p' - p_c)^{-\nu} = \xi_0(\Lambda(p - p_c))^{-\nu} = \xi_0 \Lambda^{-\nu}(p - p_c)^{-\nu} \,. \tag{7.15}$$

We can rewrite this using $\xi(p)$

$$\xi(p') = \Lambda^{-\nu}\xi(p) \,. \tag{7.16}$$

However, we also know that

$$\xi(p') = \frac{1}{b}\xi(p) \,. \tag{7.17}$$

Consequently, we have found that

$$b = \Lambda^{\nu} \,. \tag{7.18}$$

This implies that the exponent ν is a property of the fixpoint of the mapping $R(p)$. We can find ν from

$$\nu = \frac{\ln b}{\ln \Lambda} \,, \tag{7.19}$$

where we remember that $\Lambda = R'(p_c)$.

7.2   Examples

In the following we provide several examples of the application of the renormalization theory. Our renormalization procedure can be summarized in the following steps

1. Coarse-grain the system into cells of size b^d.
2. Find a rule to determine the new occupation probability, p', from the old occupation probability, p: $p' = R(p)$.
3. Determine the non-trivial fixpoints, p^*, of the renormalization mapping: $p^* = R(p^*)$, and use these points as approximations for p_c: $p_c = p^*$.
4. Determine the rescaling factor Λ from the renormalization relation at the fixpoint: $\Lambda = R'(p^*)$.
5. Find ν from the relation $\nu = \ln b / \ln \Lambda$.

It is important to realize that the renormalization mapping $R(p)$ is not unique.In order to obtain useful results we should ensure that the mapping preserves connectivity on average.

Example: One-Dimensional Percolation

Let us first address the one-dimensional percolation problem using the renormalization procedure. We have illustrated the one-dimensional percolation problem in Fig. 7.5. We generate the renormalization mapping by ensuring that it conserves connectivity. The probability for two sites to be connected over a distance b is p^b when the occupation probability for a single site is p. A renormalization mapping that conserves connectivity is therefore:

$$p' = \Pi(p, b) = p^b . \tag{7.20}$$

The fixpoints for this mapping are

$$p^* = (p^*)^b , \tag{7.21}$$

with only two possible solutions, $p^* = 0$, and $p^* = 1$. An example of a renormalization iteration is shown in Fig. 7.6. The curve illustrates that $p^* = 0$ is the only attractive or stable fixpoint, and that $p^* = 1$ is an unstable fixpoint.

Fig. 7.5 Illustration of a renormalization rule for a one-dimensional site percolation system with $b = 3$

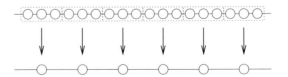

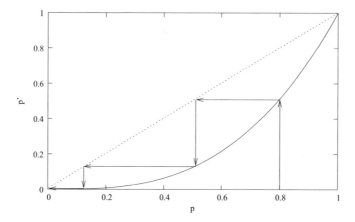

Fig. 7.6 Illustration of a renormalization rule for a one-dimensional site percolation system with $b = 3$

We can also apply the theory directly to find the exponent ν. The renormalization relation is $p' = R(p) = p^b$. We can therefore find Λ from:

$$\Lambda = \left. \frac{\partial R}{\partial p} \right|_{p^*} = b(p^*)^{b-1} = b \,, \tag{7.22}$$

where we are now studying the unstable fixpoint $p^* = 1$. We can therefore determine ν from (7.19):

$$\nu = \frac{\ln b}{\ln \Lambda} = 1 \,. \tag{7.23}$$

We notice that b was eliminated in this procedure, which is essential since we do not want the exponent to depend on details such as the size of renormalization cell. The result for the scaling of the correlation length is therefore

$$\xi \propto \frac{1}{1 - p} \,, \tag{7.24}$$

when $1 - p \ll 1$.

Example: Renormalization on 2d Site Lattice

Let us now use this method to address a renormalization scheme for two-dimensional site percolation. We will use a scheme with $b = 2$. The possible configurations for a 2×2 lattice are shown in Fig. 7.7.

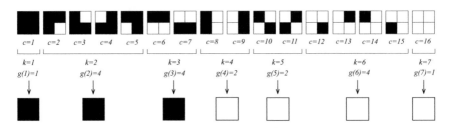

Fig. 7.7 Possible configurations for a 2×2 site percolation system. The top row indicates various configurations and the middle row how the configurations are categorized into 7 classes k, where $g(k)$ is the number of configurations in class k

In order to preserve connectivity, we need to ensure that classes $k = 1$ and $k = 2$ are occupied also in the renormalized lattice. However, we have some freedom as to which configurations to include in the class $k = 3$ and $k = 4$. We may choose only to consider spanning in one direction or spanning in both directions. In the mapping in Fig. 7.7 we only include horizontal spanning. Then the renormalization relation becomes

$$p' = R(p) = \sum_k g(k) P(k) f(k) = 1 \cdot p^4 \cdot 1 + 4 \cdot p^3 (1 - p)^1 \cdot 1$$

$$+ 2 \cdot p^2 (1 - p)^2 \cdot 1 + 2 \cdot p^2 (1 - p)^2 \cdot 0 + 2 \cdot p^2 (1 - p)^2 \cdot 0$$
$$+ 4 \cdot p^1 (1 - p)^3 \cdot 0 + 1 \cdot (1 - p)^4 \cdot 0 = p^4 + 4p^3 (1 - p) + 2p^2 (1 - p^2) .$$
$$(7.25)$$

where $f(k) = 1$ if class k is mapped onto an occupied site and $f(k) = 0$ if class k is mapped onto an empty site. The renormalization relation is illustrated in Fig. 7.8.

We will now follow steps 3 and 4. First, in step 3, we determine the fixpoints of the renormalization relation. That is, we find the solutions to the equation

$$p^* = R(p^*) = (p^*)^4 + 4(p^*)^3 (1 - p^*) + 2(p^*)^2 (1 - p^*)^2 . \qquad (7.26)$$

The trivial solution $p^* = 0$ is not of interest. Therefore we divide by p^* to produce

$$(p^*)^3 + 4(p^*)^2 (1 - p^*) + 2(p^*)(1 - p^*)^2 = 1 . \qquad (7.27)$$

The other trivial fixpoint is $p^* = 1$. We divide the equation by $1 - p^*$ to get

$$(p^*)^2 + p^* - 1 = 0 . \qquad (7.28)$$

The solutions to this second order equation are

$$p^* = -\frac{1 \pm \sqrt{1 + 4}}{2} = \frac{\sqrt{5} \pm 1}{2} \simeq 0.62 . \qquad (7.29)$$

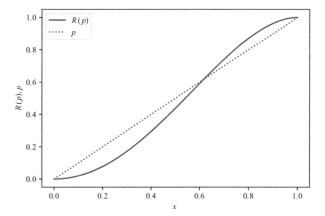

Fig. 7.8 Plot of the renormalization relation $p' = R(p) = p^4 + 4p^3(1-p) + 2p^2(1-p)^2$ for a two-dimensional site percolation problem

We have therefore found an estimate of p_c by setting $p_c = p^*$. This does not produce the correct value for p_c in a two-dimensional site percolation system, but the result is still reasonably correct. We can similarly estimate the exponent ν by calculating $R'(p^*)$.

Example: Renormalization on 2d Triangular Lattice

We will now use the same method to address percolation on site percolation on a triangular lattice. A triangular lattice is a lattice where each point has six neighbors. In solid state physics, the lattice is known as the hexagonal lattice because of its hexagonal rotation symmetry. Site percolation on the triangular lattice is particularly well suited for renormalization treatment, because a coarse grained version of the lattice is also a triangular lattice, as illustrated in Fig. 7.9, with a lattice spacing $b = \sqrt{3}$ times the original lattice size.

We will use the majority rule for the renormalization mapping. That is, we will map a set of three sites onto an occupied site if a majority of the sites are occupied, meaning that two or more sites are occupied. Otherwise, the renormalized site is empty. This mapping is illustrated in Fig. 7.9. This mapping does, as the reader may easily check, on the average conserve connectivity. The renormalization mapping becomes

$$p' = R(p) = p^3 + 3p^2(1-p) = 3p^2 - 2p^3 . \tag{7.30}$$

The fixpoints of this mapping are the solutions of the equation

$$p^* = 3(p^*)^2 - 2(p^*)^3 . \tag{7.31}$$

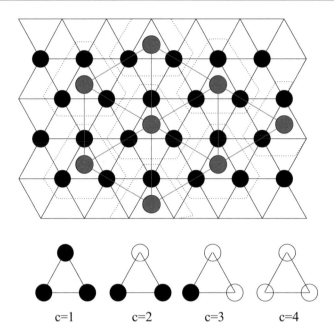

Fig. 7.9 Illustration of a renormalization scheme for site percolation on a triangular lattice. The rescaling factor is $b = \sqrt{3}$, and we use the majority rule for the mapping, that is, classes $k = 1$ and $k = 2$ are occupied, and classes $k = 3$ and $k = 4$ are mapped onto empty sites. Here, $g(1) = 1$, $g(2) = 3$, $g(3) = 3$ and $g(4) = 1$, giving $8 = 2^3$ configurations

We observe that the trivial fixpoints $p^* = 0$ and $p^* = 1$ indeed satisfy (7.31). The non-trivial fixpoint is $p^* = 1/2$. We are pleased to observe that this is the exact solution for p_c for site percolation on the triangular lattice.

We can use this relation to determine the scaling exponent ν. First, we calculate Λ:

$$\Lambda = R'(p^*) = 6p(1 - p)|_{p=\frac{1}{2}} = \frac{3}{2} \, . \tag{7.32}$$

As a result we find the exponent ν from

$$\frac{1}{\nu} = \frac{\ln \Lambda}{\ln b} = \frac{\ln 3/2}{\ln \sqrt{3}} \simeq 1.355 \, , \tag{7.33}$$

which is very close to the exact result $\nu = 4/3$ for two-dimensional percolation.

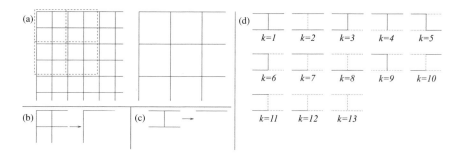

Fig. 7.10 (a) Illustration of a renormalization scheme for bond percolation on a square lattice in two dimensions. The rescaling factor is $b = 2$. (**b**) In general, the renormalization involves a mapping from 8 to two bonds. However, we will consider percolation only in the horizontal direction. This simplifies the mapping, to the figure shown in (**c**). For this mapping, the classes are shown and enumerated in (**d**)

Example: Renormalization on 2d Bond Lattice

As our last example of renormalization in two-dimensional percolation problems, we will study the bond percolation problem on a square lattice. The renormalization procedure is shown in Fig. 7.10. In the renormalization procedure, we replace 8 bonds by 2 new bonds. We consider connectivity only in the horizontal direction, and may therefore simplify the lattice, by only considering the mapping of the H-cell, a mapping of five bonds onto one bond in the horizontal direction. The various configurations are shown in the figure. In Table 7.1 we have shown the number of such configurations, and the probabilities for each configuration, which is needed in order to calculate the renormalization connection probability p'.

The resulting renormalization equation is given as

$$p' = R(p) = \Pi = \sum_{c=1}^{13} n(c) P(c) \Pi|c , \qquad (7.34)$$

where we have used k to denote the various classes, $P(k)$ is the probability for one instance of class k, $n(k)$ is the number of different configurations due to symmetry consideration in class k, and $\Pi|k$ is the spanning probability given that the configuration is in class k. The resulting relation is

$$p' = R(p) \qquad (7.35)$$

$$= p^5 + p^4(1-p) + 4p^4(1-p) + 2p^3(1-p)^2 \qquad (7.36)$$

$$+2p^3(1-p)^2 + 4p^3(1-p)^2 + 2p^2(1-p)^3 \qquad (7.37)$$

$$= 2p^5 - 5p^4 + 2p^3 + 2p^2 . \qquad (7.38)$$

Table 7.1 A list of the
possible classes k for
renormalization of a bond
lattice. The probability for
percolation given that the
class is k is denoted $\Pi|k$. The
spanning probability for the
whole cell is then $\Pi(p) =$
$p' = \sum_k n(k) P(k) \Pi|k$

k			$P(k)$	$n(k)$	$\Pi|k$
1			$p^5(1-p)^0$	1	1
2			$p^4(1-p)^1$	1	1
3			$p^4(1-p)^1$	4	1
4			$p^3(1-p)^2$	2	1
5			$p^3(1-p)^2$	2	1
6			$p^3(1-p)^2$	2	0
7			$p^3(1-p)^2$	4	1
8			$p^2(1-p)^3$	2	1
9			$p^2(1-p)^3$	4	0
10			$p^2(1-p)^3$	2	0
11			$p^2(1-p)^3$	2	0
12			$p^1(1-p)^4$	5	0
13			$p^0(1-p)^5$	1	0

The fixpoints for this mapping are $p^* = 0$, $p^* = 1$, and $p^* = 1/2$. The fixpoint
$p^* = 1/2$ provides the exact solution for the percolation threshold on the bond
lattice in two dimensions. We find Λ by derivation

$$\Lambda = R'(p^*) = \frac{13}{8} . \tag{7.39}$$

The corresponding estimate for the exponent ν is

$$\nu = \frac{\ln b}{\ln \Lambda} \simeq 1.428 , \tag{7.40}$$

which should be compared with the exact result of $\nu = 4/3$ for two-dimensional
percolation.

Exercises

Exercise 7.1 (Renormalization of nnn-Model)

(a) Develop a renormalization scheme for a two-dimensional site percolation system with next-nearest neighbor (nnn) connectivity. That is, list the 16 possible configurations, and determine what configuration they map onto in the renormalized lattice.
(b) Find the renormalized occupation probability $p' = R(p)$.
(c) Plot $R(p)$ and $f(p) = p$.
(d) Find the fixpoints p^* so that $R(p^*) = p^*$.
(e) Find the rescaling factor $\Lambda = R'(p^*)$.
(f) Determine the exponent $\nu = \ln \Lambda / \ln b$.
(g) How can we improve the estimates of p_c and ν?

Exercise 7.2 (Renormalization of Three-Dimensional Site Percolation Model)

(a) Find all 2^8 possible configurations for the $2 \times 2 \times 2$ renormalization cell for three-dimensional site percolation.
(b) Determine a renormalization scheme - what configurations map onto an occupied site?
(c) Find the renormalized occupation probability $p' = R(p)$.
(d) Plot $R(p)$ and $f(p) = p$.
(e) Find the fixpoints p^* so that $R(p^*) = p^*$.
(f) Find the rescaling factor $\Lambda = R'(p^*)$.
(g) Determine the exponent $\nu = \ln \Lambda / \ln b$.

Exercise 7.3 (Renormalization of Three-Dimensional Bond Percolation Model)
In this exercise we will develop an H-cell renormalization scheme for bond percolation in three dimensions. The three-dimensional H-cell is illustrated in Fig. 7.11.

Fig. 7.11 Illustrations of the 3d H-cell

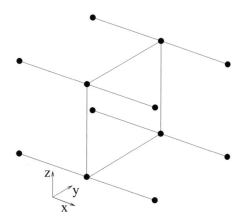

(a) Find all 2^12 possible configurations for this H-cell.
(b) Determine a renormalization scheme - what configurations map onto an occupied site?
(c) Find the renormalized occupation probability $p' = R(p)$.
(d) Plot $R(p)$ and $f(p) = p$.
(e) Find the fixpoints p^* so that $R(p^*) = p^*$.
(f) Find the rescaling factor $\Lambda = R'(p^*)$.
(g) Determine the exponent $\nu = \ln \Lambda / \ln b$.

Exercise 7.4 (Numerical Study of Renormalization) Use the following program
to study the renormalization of a given sample of a percolation system.

```
# Coarsening procedure
import numpy as np
import matplotlib.pyplot as plt
from scipy.ndimage import measurements

def coarse(z,f):
#function zz = coarse(z,f)
# The original array is z
# The transfer function is f given as a vector
#      with 16 possible places
# f applied to a two-by-two matrix should return
# the renormalized values
#
# The various values of f correspond to the following
# configurations of the two-by-two region that is renormalized,
# where I have used X to mark a present site, and 0 to mark an
# empty sites
#
#  0   00      4   00      8   00     12   00
#      00          X0          0X          XX
#
#  1   X0      5   X0      9   X0     13   X0
#      00          X0          0X          XX
#
#  2   0X      6   0X     10   0X     14   0X
#      00          X0          0X          XX
#
#  3   XX      7   XX     11   XX     15   XX
#      00          X0          0X          XX
#
    nx = np.shape(z)[0]
    ny = np.shape(z)[1]
    if (nx%2==1): # Must be even number
        raise ValueError('nx must be even')
    if (ny%2==1): # Must be even number
        raise ValueError('ny must be even')
    nx2 = int(nx/2)
    ny2 = int(ny/2)
    zz = np.zeros((nx2,ny2),float) # Generate return matrix
```

```
        x = np.zeros((2,2),float)
        for iy in range(0,ny,2):
            for ix in range(0,nx,2):
                x = z[ix,iy]*1 + z[ix,iy+1]*2 + \
                z[ix+1,iy]*4 + z[ix+1,iy+1]*8
                xx = f[int(x)]
                zz[int((ix+1)/2),int((iy+1)/2)] = xx
        return zz

# Example of use of the coarsening procedure
L = 64
m = np.random.rand(L,L)
ngen = 7
percimg = []
p = 0.65
z = (m<p)*1.0
# Set up array for transformation f
f = [0,0,0,1,0,0,0,1,0,0,0,1,1,1,1,1]
# Generate labels and loop for next
for i in range(ngen):
    lw,num = measurements.label(z)
    area = measurements.sum(z,lw,index=np.arange(lw.max()+1))
    areaImg = area[lw]
    percimg.append(areaImg)
    if (i<ngen-1): # coarse grain for next level
        zz = coarse(z,f)
        z = zz

# Plot the results
fig = plt.figure(figsize=(4*ngen,3.5))
for i in range(ngen):
    ax = fig.add_subplot(1,ngen,i+1)
    zi = percimg[i]
    ax.imshow(zi)
    ax.set_aspect('equal')
```

Perform successive iterations for $p = 0.3$, $p = 0.4$, $p = 0.5$, $p = p_c$, $p = 0.65$, $p = 0.70$, and $p = 0.75$, in order to understand the instability of the fixpoint at $p = p_c$.

Subset Geometry

8

So far, we have studied the geometry of the percolation system. Now, we will gradually address the physics of processes that occur in a percolation system. We have addressed one physics-like property of the system, the *density* of the spanning cluster, and we found that we could build a theory for the density P as a function of the porosity (occupation probability) p of the system. In order to address other physical properties, we need to have a clear description of the geometry of the percolation system close to the percolation threshold. In this chapter, we will develop a simplified geometric description that will be useful, indeed essential, when we discuss physical process in disordered media. We will introduce various subsets of the spanning cluster—sets that play roles in specific physical processes. We will start by introducing *singly connected bonds*, *the backbone* and *dangling ends* and provide a simplified image of the spanning cluster in terms of the *blob* model for the percolation system [2, 9, 16, 34].

8.1 Singly Connected Bonds

We will start with an example of a subset of the spanning cluster, the set of *singly connected sites* (or bonds). This will demonstrate what we mean by a subset and how the subset is connected to a physical problem.

> **Singly Connected Site** A singly connected site is a site with the property that if it is removed, the spanning cluster will no longer be spanning.

We can relate this to a physical property: If we study fluid flow in the spanning cluster, all the fluid has to go through the singly connected sites. These sites are also often referred to as red sites, because if we were studying a set of random

A. Malthe-Sørenssen, *Percolation Theory Using Python*, Lecture Notes in Physics 1029, https://doi.org/10.1007/978-3-031-59900-2_8

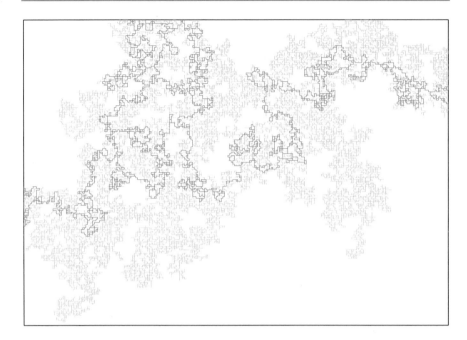

Fig. 8.1 Illustration of the spanning cluster, the singly connected bonds (red), the backbone (blue), and the dangling ends (green) for a 256 × 256 bond percolation system at $p = p_c$. (Figure from Martin Søreng)

resistors, the highest current would have to go through the singly connected bonds, and they would therefore heat up and become "red". Several examples of subsets of the spanning cluster, including the singly connected bonds, are shown in Fig. 8.1.

Scaling Hypothesis We have learned that the spanning cluster may be described by the mass scaling relation $M \propto L^D$, where D is termed the fractal dimension of the spanning cluster. Here, we will make a daring hypothesis, which we will also substantiate: We propose that subsets of the spanning cluster obey similar scaling relations.

For example, we propose that the mass of the singly connected sites (M_{SC}) has the scaling form

$$M_{SC} \propto L^{D_{SC}} , \tag{8.1}$$

where we call the dimension D_{SC} the fractal dimension of the singly connected sites. Because the set of singly connected sites is a subset of the spanning cluster, we know that $M_{SC} \leq M$. It therefore follows that

$$D_{SC} \leq D . \tag{8.2}$$

Based on this simple example, we will generalize the approach to other subsets of the spanning cluster. However, first we will introduce a new concept, a self-avoiding path on the spanning cluster.

8.2 Self-Avoiding Paths on the Cluster

The study of percolation is the study of connectivity, and many of the physical properties that we are interested in depends on various forms of connecting paths on the spanning cluster between two opposite edges. We can address the structure of connected paths between the edges by studying self-avoiding paths (SAPs). A Self-Avoiding Path (SAP) is a set of connected sites that correspond to the sites on the path of a walk on the spanning cluster that does not intersect itself going from one side to the opposite side.

Minimal Path

The shortest path between the two edges is called the shortest SAP between the two edges. (Notice, that there may be more than one path the satisfy this criterion. We chose one of these paths randomly). We call this the *minimal path* and denote its length $L_{\min}$. The length here refers to the number of sites in the path, which we also call the mass of the path, $M_{\min} = L_{\min}$. We will use mass instead of length in the following to describe the paths.

We assume that mass of the minimal path also scales with the system size according to the scaling form:

$$M_{\min} \propto L^{D_{\min}} . \tag{8.3}$$

Where we have introduced the scaling exponent of the minimal path to be $D_{\min}$.

Maximum and Average Path

Similarly, we call the longest SAP between the two edges the *longest path* with a mass $M_{\max}$. Again, we assume that the mass has a scaling form, $M_{\max} \propto L^{D_{\max}}$. We notice that $M_{min} \leq M_{max}$. Consequently, a similar relation holds for the exponents $D_{\min} \leq D_{\max}$.

We also introduce the term the *average path*, meaning the average mass (length) of all possible SAPs going between opposite sides of the system, $\langle M_{SAP} \rangle \propto L^{D_{SAP}}$. The dimension D_{SAP} will lie between the dimensions of the minimal and the maximal path.

Backbone

Intersection of All Self-Avoiding Paths The notion of SAPs can also be used to address the physical properties of the cluster, such as we saw for the singly connected bonds. The set of singly connected bonds is the set of intersections between all SAPs connecting the two sides. That is, the singly connected bonds is the set of points that any path must go through in order to connect the two sides. From this definition, we notice that the dimension $D_{SC} < D_{min}$, and as we will see further on, $D_{SC} = 1/\nu$ which is smaller than 1 for two-dimensional systems.

Union of All Self-Avoiding Paths Another useful set is the union of all SAPs that connect the two edges of the cluster. This set is called the backbone with dimension D_B.

> **Backbone** The *backbone* is the union of all self-avoiding paths on the spanning cluster that connect two opposite edges.

This set has a simple physical interpretation for a random porous material, since it corresponds to the sites that are accessible to fluid flow if a pressure is applied across the material. The remaining sites are called *dangling ends*. The backbone are all the sites that have at least two different paths leading into them, one path from each side of the cluster. The remaining sites only have one (self-avoiding) path leading into them, and we call this set of sites the dangling ends. The spanning cluster consists of the backbone plus the dangling ends, as illustrated in Fig. 8.2. The dangling ends are therefore pieces of the cluster that can be cut away by the removal of a single bond.

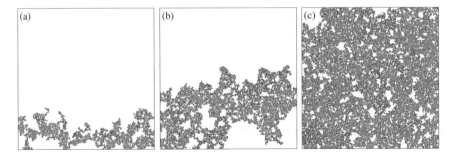

Fig. 8.2 Illustration of the spanning cluster consisting of the backbone (red) and the dangling ends (blue) for a 512×512 site percolation system for (**a**) $p = 0.58$, (**b**) $p = 0.59$, and (**c**) $p = 0.61$

We have arrived at the following hierarchy of exponents describing various subsets of paths through the cluster:

$$D_{SC} \leq D_{\min} \leq D_{SAP} \leq D_{\max} \leq D_B \leq D \leq d , \qquad (8.4)$$

Scaling of the Dangling Ends

Generally, we will find that the dimension of the backbone, D_B, is smaller than the dimension of the spanning cluster. For example, in two dimensions, we find that $D_B \simeq 1.6$, whereas $D \simeq 1.89$. This has implications for the relative size of the backbone and the dangling ends.

The spanning cluster consists of the backbone and the dangling ends. Therefore, the mass of the spanning cluster, M, must equal the sum of the masses of the backbone and the dangling ends $M = M_B + M_{DE}$. Since we know that $M \propto L^D$ and $M_B \propto L^{D_B}$, we find that

$$M_{DE} = M - M_B = M_0 L^D - M_{0,B} L^{D_B} , \qquad (8.5)$$

where M_0 and $M_{0,B}$ are constant prefactors. To see what happens when $L \to \infty$, we divide by M:

$$\frac{M_{DE}}{M} = 1 - \frac{M_{0,B} L^{D_B}}{M_0 L^D} = 1 - c L^{D_B - D} , \qquad (8.6)$$

Since $D_B \leq D$, we see that the fraction M_{DE}/M goes to a constant (one) as L approaches infinity. Consequently, we have found that $M_{DE} \propto M \propto L^D$. This also implies that as the system size goes to infinity most of the mass is in the dangling ends. This means that the backbone occupies a smaller and smaller portion of the total mass of the system as the system size increases.

Argument for the Scaling of Subsets

We can provide a better argument for why the various subsets should scale with the system size L to various exponents. We notice that the following relation between the masses must be true:

$$L^1 \leq M_{\min} \leq M_{SAP} \leq M_{\max} \leq M_{BB} \leq M \leq L^d , \qquad (8.7)$$

where the first inequality $L^1 \leq M_{\min}$ follows because even the minimum path must be at least of length L to go from one side to the opposite side.

Now, if this is to be true for all values of L, it can be argued that because all the masses are between two scaling relations, L^1 and L^d, also the scaling of the intermediate masses, M_x, must be power-laws with some power-law exponents, $M_x \propto L^{D_x}$, with the hierarchy of exponents given in (8.4).

Blob Model for the Spanning Cluster

Let us now try to formulate our geometric description of the spanning cluster into a *model* of the spanning cluster [36]. We have found that the spanning cluster can be subdivided first into two parts: the backbone and the dangling ends. The backbone may again be divided into two parts: a set of blobs where the are several parallel paths and a set of sites, the singly connected sites, that connect the blobs to each other and the blobs to the dangling ends. Thus, we have ended up with a model with three components:

- the dangling ends,
- a set of blobs where there are several parallel paths
- the singly connected points, connecting the blobs to each other and the blobs to the dangling ends.

Each of the blobs and the dangling ends will again have a similar substructure of dangling ends, blobs with parallel paths, and singly connected bonds as illustrated in Fig. 8.3. This cartoon image of the clusters provides very useful intuition about the geometrical structure of percolation clusters, which we will use when we address the physics of disordered systems in the next chapters.

Mass-Scaling Exponents for Subsets of the Spanning Clusters

The exponents can be calculated either by numerical simulations, where the masses of the various subsets are measured as a function of system size at $p = p_c$, or by the renormalization group method. Numerical results based on computer simulations using the code provided in this book are listed in Table 8.1. You can find up-to-date

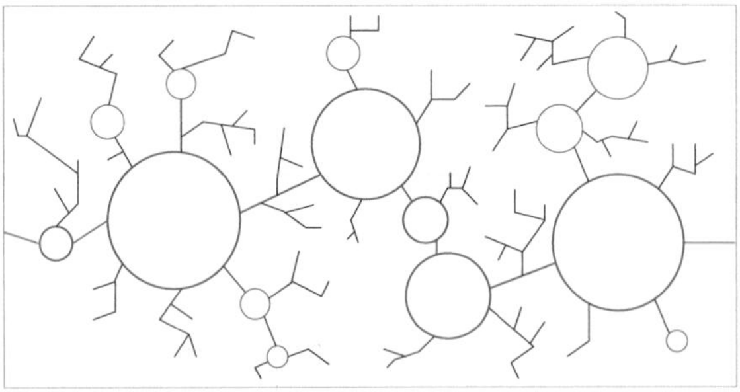

Fig. 8.3 Illustration of the hierarchical blob-model for the percolation cluster showing the backbone (bold), singly connected bonds (red) and blobs (blue)

Table 8.1 A list of known exponent for the various subset types in two dimensions

d	D_{SC}	$D_{\min}$	$D_{\max}$	D_B	D	D_{DE}
2	0.75	1.1	1.5	1.6	1.89	1.89

results for exponents in the percolation system at the Wikipedia page: https://en. wikipedia.org/wiki/Percolation_critical_exponents.

8.3 Renormalization Calculation

We will now use the renormalization group approach to address the scaling exponent for various subsets of the spanning cluster at $p = p_c$. For this, we will here use the renormalization procedure for bond percolation on a square lattice in two dimensions following Hong and Stanley [17], where we have found that the renormalization procedure produces the exact result for the percolation threshold, $p_c = p^* = 1/2$, which is a fixpoint of the mapping.

Our strategy will be to assume that all the bonds have a mass $M = 1$ in the original lattice, and then find the mass M' in the renormalized lattice, when the length has been rescaled by b. For a property that displays a self-similar scaling, we will expect that

$$M' \propto b^{D_x} M , \tag{8.8}$$

where D_x denotes the dimension for the particular subset we are looking at. We can use this to determine the fractal exponent D_x from

$$D_x = \frac{\ln M'/M}{\ln b} . \tag{8.9}$$

We will do this by calculating the average value of the mass of the H-cell, by taking the mass of the subset we are interested in for each configuration, $M_x(c)$, and multiplying it by the probability of that configuration, summing over all configurations:

$$\langle M \rangle = \sum_c M_x(c) P(c) . \tag{8.10}$$

We have now calculated the average mass in the original 2 by 2 lattice, and this should correspond to the average renormalized mass, $\langle M' \rangle = p'M'$, which is the mass of the renormalized bond, M' multiplied with the probability for that bond to be present p'. That is, we will find M' from:

$$p'M' = \sum_c M(c) P(c) , \tag{8.11}$$

Table 8.2 Numerical exponents for the exponent describing various subsets of the spanning cluster defined using the set of Self-Avoiding Walks going from one side to the opposite side of the cluster. The last line shows the exponents found from numerical simulations in a two-dimensional system

c		$P(c)$	M_{SC}	L_{min}	L_{AVG}	L_{max}	M_{BB}	M
1		$p^5(1-p)^0$	0	2	2.5	3	5	5
2		$p^4(1-p)^1$	0	2	2	2	4	4
3		$4p^4(1-p)^1$	1	2	2.5	3	4	4
4		$2p^3(1-p)^2$	2	2	2	2	2	3
5		$2p^3(1-p)^2$	3	3	3	3	3	3
6		$4p^3(1-p)^2$	2	2	2	2	2	3
7		$2p^2(1-p)^3$	2	2	2	2	2	2
$\langle M_x \rangle$			$26/2^5$	$34/2^5$	$36.5/2^5$	$39/2^5$	$47/2^5$	$53/2^5$
D_x			$\frac{\ln \frac{13}{8}}{\ln 2}$	$\frac{\ln \frac{17}{8}}{\ln 2}$	$\frac{\ln \frac{36.5}{16}}{\ln 2}$	$\frac{\ln \frac{39}{16}}{\ln 2}$	$\frac{\ln \frac{47}{16}}{\ln 2}$	$\frac{\ln \frac{53}{16}}{\ln 2}$
D_x			0.7004	1.0875	1.1898	1.2854	1.5546	1.7279
$D_{x,n}$			3/4	1.13		1.4	1.6	91/48

We will study our system at the nontrivial fixpoint $p = p^* = 1/2 = p_c$. The spanning configurations c for bond renormalization in two dimensions, are shown together with their probabilities and the masses of various subsets in Table 8.2.

This use of the renormalization group method to estimate the exponents demonstrates the power of the renormalization arguments. Similar arguments will be used to address other properties of the percolation system.

8.4 Deterministic Fractal Models

We have found that we can calculate the behavior of infinite-dimensional and one-dimensional systems exactly. However, for finite dimensions such as for $d = 2$ or $d = 3$ we must rely on numerical simulations and renormalization group arguments to determine the exponents and the behavior of the system. However, in order to learn about physical properties in systems with scaling behavior, we may be able to construct simpler models that contain many of the important features of the percolation cluster. For example, we may be able to introduce deterministic, iterative fractal structures that reproduce many of the important properties of the percolation cluster at $p = p_c$, but that are deterministic systems. The idea is that we can use such a system to study other properties of the physics on fractal structures.

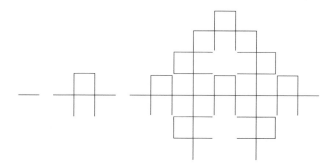

Fig. 8.4 Illustration of first three generations of the Mandelbrot-Given curve. The length is scaled by a factor $b = 3$ for each iteration, and the mass of the whole structure is increased by a factor of 8. The fractal dimension is therefore $D = \ln 8/\ln 3 \simeq 1.89$

Mandelbrot-Given Curve An example of an iterative fractal structure that has many of the important features of the percolation clusters at $p = p_c$ is the Mandelbrot-Given curve. The curve is generated by the iterative procedure described in Fig. 8.4. Through each generation, the length is rescaled by a factor $b = 3$, and the mass is rescaled by a factor 8. That is, for generation l, the mass is $m(l) = 8^l$, and the linear size of the cluster is $L(l) = 3^l$. If we assume a scaling on the form $m = L^D$, we find that

$$D = \frac{\ln 8}{\ln 3} \simeq 1.89 \ . \tag{8.12}$$

This is surprisingly similar to the fractal dimension of the percolation cluster. We can also look at other dimensions, such as for the singly connected bonds, the minimum path, the maximum path and the backbone.

Single Connected Bonds Let us first address the singly connected bonds. In the zero'th generation, the system is simply a single bond, and the length of the singly connected bonds, L_{SC} is 1. In the first generation, there are two bonds that are singly connecting, and in the second generation there are four bonds that are singly connecting. The general relation is that

$$L_{SC} = 2^l \ , \tag{8.13}$$

where l is the generation of the structure. The dimension, D_{SC}, of the singly connected bonds is therefore $D_{SC} = \ln 2/\ln 3 \simeq 0.63$, which should be compared with the exact value $D_{SC} = 3/4$ for two-dimensional percolation.

Minimum Path The minimum path will for all generations be the path going straight through the structure, and the length of the minimal path will therefore be equal to the length of the structure. The scaling dimension D_{min} is therefore $D_{min} = 1$.

Maximum Path   The maximum path increases by a factor 5 for each iteration. The dimension of the maximum path is therefore $D_{max} = \ln 5/\ln 3 \simeq 1.465$.

Backbone   We can similarly find that the mass of the backbone increases by a factor 6 for each iteration, and the dimension of the backbone is therefore $D_B = \ln 6/\ln 3 \simeq 1.631$.

Model System   This deterministic iterative fractal can be used to perform quick calculations of various properties on a fractal system, and may also serve as a useful hierarchical lattice on which to perform simulations when we are studying processes occurring on a fractal structure.

8.5   Lacunarity

The fractal dimension describes the scaling properties of structures such as the percolation cluster at $p = p_c$. However, structures that have the same fractal dimension, may have a very different appearance. As an example, let us study several variations of the Sierpinski gasket introduced in Sect. 5.3. As illustrated in Fig. 8.5, we can construct several rules for the iterative generation of the fractal that all result in the same fractal dimension, but have different visual appearance. The fractal dimension $D = \ln 3/\ln 2$ for both of the examples in Fig. 8.5, but by increasing the number of triangles that are used in each generation, the structures become more homogeneous. How can we quantify this difference?

Distribution of Mass   In order to quantify this difference, Mandelbrot invented the concept of lacunarity . We measure lacunarity from the distribution of mass-

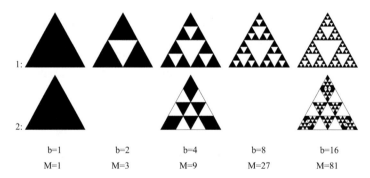

Fig. 8.5 Two versions of the Siepinski gasket. In version 1, the next generation is made from 3 of the structures from the last generation, and the spatial rescaling is by a factor $b = 3$. In version 2, the next generation is made from 9 of the structures from the last generation, and the spatial rescaling is by a factor $b = 6$. The resulting fractal dimension is $D_2 = \ln 9/\ln 4 = \ln 3^2/\ln 2^2 = \ln 3/\ln 2 = D_1$. The two structures therefore have the same fractal dimension. However, version 1 have large fluctuations that version 2

sizes. We can characterize and measure the fractal dimension of a fractal structure using box-counting, as explained in Sect. 5.3. The structure, such as the percolation cluster, is divided into boxes of size ℓ. In each box, i, there will be a mass $m_i(\ell)$. The fractal dimension is found by calculating the average mass per box of size ℓ:

$$\langle m_i(\ell) \rangle_i = A\ell^D . \tag{8.14}$$

However, there will variations in the masses $m(\ell)$ in the boxes, characterized by a distribution $P(m, \ell)$, which gives the probability for mass m in a box of size ℓ. We can characterize this distribution by its moments:

$$\langle m^k(\ell) \rangle = A_k \ell^{kD} , \tag{8.15}$$

where this particular scaling form implies that the structure is unifractal: the scaling exponents for all the moments are linearly related.

Unifractal Scaling For a unifractal structure, we expect the distribution of masses to have the scaling form

$$P(m, \ell) = \ell^x f\left(\frac{m}{\ell^D}\right) , \tag{8.16}$$

where the scaling exponent x is yet undetermined. In this case, the moments can be found by integration over the probability density

$$\langle m^k \rangle = \int P(m, \ell) m^k \, dm \tag{8.17}$$

$$= \int m^k \ell^x f\left(\frac{m}{\ell^D}\right) dm \tag{8.18}$$

$$= \ell^{(kD+x+D)} \int \left(\frac{m}{\ell^D}\right)^k f\left(\frac{m}{\ell^D}\right) d\left(\frac{m}{\ell^D}\right) \tag{8.19}$$

$$= \ell^{D(k+1)+x} \int x^k f(x) \, dx \tag{8.20}$$

We can determine the unknown scaling exponent x from the scaling of the zero'th moment, that is, from the normalization of the probability density: $\langle m^0 \rangle = 1$ implies that $D(0 + 1) + x = 0$, and therefore, that $x = -D$. The scaling ansatz for the distribution of masses is therefore

$$P(m, \ell) = \ell^{-D} f\left(\frac{m}{\ell^D}\right) . \tag{8.21}$$

And we found that the moments can be written as

$$\langle m^k \rangle = \ell^{D(k+1)-D} \int x^k f(x) dx = A_k \ell^{kD} , \tag{8.22}$$

as we assumed above. Consequently, the distribution of masses is characterized by the distribution $P(m, \ell)$, which in turn is described by the fractal dimension, D, and the scaling function $f(u)$, which gives the shape of the distribution.

Properties of the Distribution of Masses The distribution of masses can be broad, which would correspond to "large holes", or narrow, which would correspond to a more uniform distribution of mass. The width of the distribution can be characterized by the mean-square deviation of the mass from the average mass:

$$\Delta = \frac{\langle m^2 \rangle - \langle m \rangle^2}{\langle m \rangle^2} = \frac{A_2 - A_1^2}{A_1^2} . \tag{8.23}$$

This number describes another part of the mass distribution relation than the scaling relation, and can be used to characterize fractal set. For the percolation problem, this number is assumed to be universal, independent of lattice type, but dependent on the embedding dimensionality of the system.

Exercises

Exercise 8.1 (Singly Connected Bonds) Use the example programs from the text to find the singly connected bonds.

(a) Run the programs to visualize the singly connected bonds. Can you understand how this algorithms finds the singly connected bonds? Why are some of the bonds of a different color?

(b) Find the mass, M_{SC}, of the singly connected bonds as a function of system size L for $p = p_c$ and use this to estimate the exponent D_{SC}: $M_{SC} \propto L^{D_{SC}}$.

(c) Can you find the behavior of $P_{SC} = M_{SC}/L^d$ as a function of $p - p_c$?

Exercise 8.2 (Left/Right-Turning Walker) We have provided a subroutine and an example program that implements the left/right-turning walker algorithm. The algorithm works on a given clusters. From one end of the cluster, two walkers are started. The walkers can only walk according to the connectivity rules on the lattice. That is, for a nearest-neighbor lattice, they can only walk to their nearest neighbors. The left-turning walker always tries to turn left from its previous direction. If this site is empty, it tries the next-best site, which is to continue straight ahead. If that is empty, it tries to move right, and if that is empty, it moves back along the direction it came. The right-turning walker follows a similar rule, but prefers to turn right in each step. The first walker to reach the other end of the cluster stops, and the other walker stops when it reaches this site.

The path of the two walkers is illustrated in the Fig. 8.6. The sites that are visited by both walkers constitute the singly connected bonds. The union of the two walks constitutes what is called the external perimeter (Hull) of the cluster.

Fig. 8.6 Illustrations of the
left-right turning walker

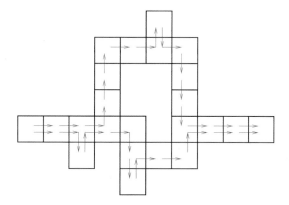

(a) Use the following programs to generate and illustrate of the singly connected
bonds for a 100×100 system. Check that the illustrated bonds correspond to
the singly connected bonds.

```python
import numpy as np
import numba

@numba.njit(cache=True)
def walk(z):
    # Left turning walker
    # Returns left: nr of times walker passes a site
    # First, ensure that array only has one contact point at
    #    left and right : topmost points chosen
    nx = z.shape[0]
    ny = z.shape[1]
    i = np.where(z[0,:] > 0)
    ix0 = 0 # starting row for walker is always 0
    iy0 = i[0][0] # starting column (first element where
    #               there is a matching column which is zero)
    # First do left-turning walker
    directions = np.zeros((4,2))
    directions [0,0] = -1 # west
    directions [0,1] = 0
    directions [1,0] = 0 # south
    directions [1,1] = -1
    directions [2,0] = 1 # east
    directions [2,1] = 0
    directions [3,0] = 0 # north
    directions [3,1] = 1
    nwalk = 1
    ix = ix0
    iy = iy0
    direction = 0 # 0 = west, 1 = south, 2 = east, 3 = north
    left = np.zeros((nx,ny))
    right = np.zeros((nx,ny))
    while (nwalk >0):
```

```python
        left[ix,iy] = left[ix,iy] + 1
        # Turn left until you find an occupied site
        nfound = 0
        while (nfound==0):
            direction = direction - 1
            if (direction < 0):
                direction = direction + 4
            # Check this direction
            iix = ix + int(directions[direction,0])
            iiy = iy + int(directions[direction,1])
            if (iix >= nx):
                nwalk = 0 # Walker escaped
                nfound = 1
                iix = nx
                ix1 = ix
                iy1 = iy
            # Is there a site here?
            elif(iix >= 0):
                if(iiy >= 0):
                    if (iiy < ny):
                        if (z[iix,iiy]>0): # site present
                            ix = iix
                            iy = iiy
                            nfound = 1
                            direction = direction + 2
                            if (direction > 3):
                                direction = direction - 4
#left
nwalk = 1
ix = ix0
iy = iy0
direction = 1 # 1=left, 2 = down, 3 = right, 4 = up
while(nwalk >0):
    right[ix,iy] = right[ix,iy] + 1
    # ix,iy
    # Turn right until you find an occupied site
    nfound = 0
    while (nfound==0):
        direction = direction + 1
        if (direction > 3):
            direction = direction - 4
        # Check this directionection
        iix = ix + int(directions[direction,0])
        iiy = iy + int(directions[direction,1])
        if (iix >= nx):
            if (iy >= iy1):
                nwalk = 0 # Walker escaped
                nfound = 1
                iix = nx
        # Is there a site here?
        elif(iix >= 0):
            if(iiy >= 0):
                if (iiy < ny):
```

```
                                 if (iix < nx):
                                     if (z[iix,iiy]>0): # site present
                                         ix = iix
                                         iy = iiy
                                         nfound = 1
                                         direction = direction - 2
                                         if (direction <0):
                                             direction = direction + 4
            return left, right


from scipy.ndimage import measurements
# Generate spanning cluster (1-r spanning)
lx = 200
ly = 200
p = 0.595
ncount = 0
perc = []
while (len(perc)==0):
    ncount = ncount + 1
    if (ncount >1000):
        print("Couldn't make percolation cluster...")
        break
    z=np.random.rand(lx,ly)<p
    lw,num = measurements.label(z)
    perc_x = np.intersect1d(lw[0,:],lw[-1,:]) # Percolating?
    perc = perc_x[np.where(perc_x > 0)]
print("ncount = ",ncount)

import matplotlib.pyplot as plt
if len(perc) > 0:
    zz = (lw == perc[0])
    # zz now contains the spanning cluster
    plt.figure(figsize=(15,8)) # Display spanning cluster
    plt.subplot(2,3,1)
    plt.imshow(zz, interpolation='nearest', origin='lower')
    l,r = walk(zz)
    plt.subplot(2,3,2)
    plt.imshow(l, interpolation='nearest', origin='lower')
    plt.subplot(2,3,3)
    plt.imshow(r, interpolation='nearest', origin='lower')
    plt.subplot(2,3,4)
    zzz = l*r # Find points where both l and r are non-zero
    plt.imshow(zzz, interpolation='nearest', origin='lower')
    plt.subplot(2,3,5)
    zadd = zz + zzz
    plt.imshow(zadd, interpolation='nearest', origin='lower')
```

(b) Measure the dimension D_{SC}.

(c) Modify the programs to find the external perimeter (Hull) of a spanning cluster in a 100×100 system.

(d) Measure the dimension D_P of the perimeter.

(e) (Advanced) Develop a theory for the behavior of $P_H(p, L)$, the probability for a site to belong to the Hull as a function of p and L for $p > p_c$.

(f) (Advanced) Measure the behavior of $P_H(p, L)$ as a function of p for $L = 512 \times 512$.

Flow in Disordered Media

9

In this chapter, we introduce the basic concepts of disordered media. We introduce properties of flow of current or fluids, and then address flow in a percolating system close to p_c. We will study the behavior numerically, develop a scaling theory, and find properties using the renormalization group approach. Our initial studies will be on the binary porous medium of the percolation system. However, we can also extend our results to more general random media, and we demonstrate how this can be done towards the end of the chapter.

9.1   Introduction to Disorder

We have now developed the tools to address the statistical properties of the geometry of a disordered system such as a model porous medium: the percolation system. In the following chapters, we will apply this knowledge to address physical properties of disordered systems and to study physical processes in disordered materials.

We have learned that the geometry of a disordered system displays fractal scaling close to the percolation threshold. Material properties such as the density of singly connected sites, or the backbone of the percolation cluster, display self-similar scaling. The backbone is the part of the spanning cluster that participates in fluid flow. The mass, M_B, of the backbone scales with the system size, L, according to the scaling relation $M_B = L^{D_B}$, where D_B is smaller than the Euclidean dimension. The density of the backbone therefore decreases with system size. This implies that material properties which we ordinarily would treat as material constants, depend on the size of the sample. In this part we will develop an understanding of the origin of this behavior, and show how we can use the tools from percolation theory to address the behavior in such systems.

The behavior of a disordered system can in principle always be addressed by direct numerical simulation. For example, for incompressible, single-phase fluid flow through a porous material, the effective permeability of a sample can be found

© The Author(s) 2024   135
A. Malthe-Sørenssen, *Percolation Theory Using Python*, Lecture Notes
in Physics 1029, https://doi.org/10.1007/978-3-031-59900-2_9

to very good accuracy from a detailed numerical model of fluid flow through the system. However, it is not practical to model fluid flow down to the smallest scale in more applied problems. We would therefore need to extrapolate from the small scale to the large scaling. This process, often referred to as up-scaling, requires that we know the scaling properties of our system. We will address up-scaling in detail in the following chapters.

We may argue that a system at the percolation threshold is anomalous and that any realistic system, such as a geological system, would be far away from the percolation threshold. In this case, the system will only display an anomalous, size-dependent behavior up to the correlation length, and over larger lengths the behavior will be that of a homogeneous material. We should, however, be aware that many physical properties are described by broad distributions of material properties, and this will lead to a behavior similar to the behavior close to the percolation threshold, as we will discuss in detail in this part. In addition, several physical processes ensure that the system is driven into or is exactly at the percolation threshold. One such example is the invasion-percolation process, which gives a reasonable description of oil-water emplacement processes such as secondary oil migration. For such systems, the behavior is well described by the scaling theory we have developed.

In this and following chapters, we will first provide an introduction to the scaling of material properties such as conductivity (Chap. 9), elasticity (Chap. 10) and diffusion (Chap. 11). Then we will demonstrate how processes occurring in systems with frozen disorder, such as a porous material, often lead to the formation of fractal structures (Chap. 12).

9.2   Conductivity and Permeability

We will start our studies of physics in disordered media by addressing flow, either in the form of incompressible fluid flow in a random, porous system or in the form of electric current in a random, porous materials. First, let us address the similarities between these two flow phenomena.

Electrical Conductivity and Resistor Networks

Traditionally, the   conductive properties of a disordered material have been addressed by studying the behavior of random networks of resistors called *random resistor networks* [1,23,24]. In this case, a voltage V is applied across the disordered material, such as a bond-percolation network, and the total current, I, through the sample is measured, giving the conductance G of the sample as the constant of proportionality $I = GV$. (We recall that the current I is the amount of charge flowing through a given cross-sectional area per unit time).

We remember from electromagnetism that we discern between *conductance* and *conductivity*:

- conductance, G, is a property of a specific sample—a given medium—with specific dimensions
- conductivity, g, is a material property

For an L^d sample in a d-dimensional system, the conductance of a homogeneous material with conductivity g is

$$G = L^{d-1}g/L = L^{d-2}g .$$
(9.1)

It is common in electromagnetism to use σ for conductivity. Here, we will instead use g to avoid confusion with the exponent σ, which we introduced previously for the behavior of s_ξ. The conductance is inversely proportional to the length of the sample in the direction of flow, and proportional to the cross-sectional $d - 1$-dimensional area. We can understand this by considering that there are L^{d-1} parallel parts that contribute to the flow. Parallel-parts add to the conductance. In addition, each part has a length L, and we recall from electromagnetism that resistance increases with length and therefore conductance decreases with length.

Flow Conductivity of a Porous System

We can also use fluid flow in porous medium as our basic physical system. If done in the limit of slow, incompressible fluid flow these two systems are practically identical. For fluid flow in a porous medium of length L and cross-sectional area A, the system is described by Darcy's law which provide a relation between, Φ, the amount of fluid volume flowing through a given cross-sectional area, A, per unit time and the pressure drop Δp across the sample:

$$\Phi = \frac{kA}{\eta}\frac{\Delta p}{L} ,$$
(9.2)

where k is the called permeability of the material and is a property of the material geometry, and η is the viscosity of the fluid. Again, we would like a description so that k is material property, and all the information about the geometry of the material goes into the permeability of the sample through the length L and the cross-sectional area A. Generalized to a d-dimensional system, the relation is

$$\Phi = \frac{kL^{d-1}}{\eta L}\Delta p = L^{d-2}\frac{k}{\eta}\Delta p .$$
(9.3)

From this, we see that the electric conductivity problem in this limit is the same as the Darcy-flow permeability problem, where $\Delta p/L$ corresponds to the voltage difference, V, and k/η corresponds to the conductivity g. We will therefore not discern between the two problems in the following. We will simply call them flow problems and describe them using the current, I, the conductivity, g, the conductance G, and the potential V. We will study these problems on a L^d percolation lattice, using the theoretical, conceptual and computational tools we have developed so far.

9.3   Conductance of a Percolation Lattice

Let us first address the conductance of a L^d percolation system. The system may be either a site or a bond percolation system, but many of the concepts we introduce are simpler to explain if we just consider a bond percolation system.

We will start with a simplified system: a network of bonds that are present with probability p. We assume that all bonds have the same conductance, which we can set to 1 without loss of generality. The bonds are removed with probability $1 - p$, and we model this by setting the conductance of a removed bond to be zero.

Finding the Conductance of the System

The conductance of the L^d sample is found by solving the flow problem illustrated in Fig. 9.1. A potential difference V is applied across the whole sample, and we find (measure) the resulting current I. We find the conductance from Ohm's law (or

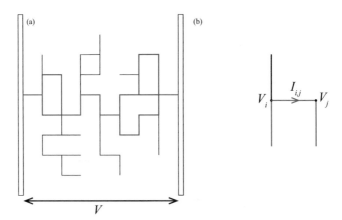

Fig. 9.1 (a) Illustration of flow through a bond percolation system. The bonds shown in red are the singly connected bonds: all the flux has to go through these bonds. The bonds shown in blue are the rest of the backbone: The flow only takes place on the singly connected bonds and the backbone, the remaining bonds are the dangling ends, which do not participate in fluid flow. (b) Illustration of the potentials V_i and V_j in two adjacent sites and the current $I_{i,j}$ from site i into site j

similarly from Darcy's law for fluid flow):

$$I = GV \quad \Rightarrow \quad G = \frac{I}{V} . \tag{9.4}$$

In general, the conductance G will be a function of p and L: $G = G(p, L)$.

Local Potentials and Currents Let us look at the network in more detail. Figure 9.1b illustrates a small part of the whole system. The two adjacent sites i and j are connected with a bond of conductance $G_{i,j}$. If the bond is present (with probability p in the percolation system), the conductance is $G_{i,j}$ is 1, otherwise it is zero.

The current from site i to site j is related to the difference in potential between the two sites:

$$I_{i,j} = G_{i,j} \left(V_i - V_j \right) , \tag{9.5}$$

where we notice that the current is positive if the potential is higher in site i than in site j.

Conservation of Current In addition, the continuity condition provides a conservation equation for the currents: The net charge (or fluid mass for Darcy flow) are conserved, and therefore the net current into any point inside the lattice must be zero. This corresponds to the condition that the sum of the current from site i into all its neighboring sites k must be zero:

$$\sum_k I_{i,k} = 0 \tag{9.6}$$

In electromagnetism this is called Kirchhoff's rule for currents. We can rewrite this in terms of the local potentials V_i instead by inserting (9.5) in (9.6):

$$\sum_k G_{i,k} \left(V_i - V_k \right) = 0 . \tag{9.7}$$

This provides us with a set of equations for all the potentials V_i, which we must solve to find the potentials and hence the currents between all the sites in a percolation system.

Finding Currents and Potentials We can use this to find all the potentials for a percolation system. Let us address a two-dimensional system of size $L \times L$. The potential in a position (x, y) on the lattice is $V(x, y)$, where x and y are integers, $x = 0, 1, 2, \ldots, L - 1$ and $y = 0, 1, 2, \ldots, L - 1$. We denote $G_{i,j}$ as $G(x_i, y_i; x_j, y_j)$. We can then rewrite (9.7) as

$$G(x, y; x + 1, y) \, (V(x, y) - V(x + 1, y)) + \tag{9.8}$$

$$G(x, y; x - 1, y) \, (V(x, y) - V(x - 1, y)) + \tag{9.9}$$

$$G(x, y; x, y + 1) \, (V(x, y) - V(x, y + 1)) + \tag{9.10}$$

$$G(x, y; x, y - 1) \, (V(x, y) - V(x, y - 1)) = 0 \tag{9.11}$$

In order to solve this two-dimensional problem, it is common to rewrite it as a one-dimensional system of equations with a single index. The index $i = x + yL$ uniquely describes a point so that $V(x, y) = V_i$. We see that $(x, y) = i$, $(x + 1, y) = i + 1$, $(x - 1, y) = i - 1$, $(x, y + 1) = i + L$, and $(x, y - 1) = i - L$. We can rewrite (9.11) using this indexing system:

$$G_{i,i+1} \, (V_i - V_{i+1}) + G_{i,i-1} \, (V_i - V_{i-1}) + \tag{9.12}$$

$$G_{i,i+L} \, (V_i - V_{i+L}) + G_{i,i-L} \, (V_i - V_{i+L}) = 0 \tag{9.13}$$

This is effectively a set of L^d equations for V_i. In addition we have the boundary conditions that $V(0, j) = V$ and $V(L - 1, j) = 0$ for $j = 0, 1, \ldots, L - 1$. This defines the system as a tri-diagonal set of linear equations that can be solved easily numerically.

Computational Methods

We have now reformulated the conductivity problem on a percolation lattice into a computational problem that we can solve. We do this by generating random lattices of size $L \times L$, solve to find the potential $V(x, y)$, and then study the effective conductivity, $G = I/V$ of the system by summing up all the currents exiting the system (or entering—these should be the same).

We can do this by generating a bond-lattice, where the values $G_{i,j}$ are either 0 or 1. However, so far all our visualization methods have been constructed for site lattices. We will therefore study a site lattice, but instead generate $G_{i,j}$ between two sites based on whether the sites are present. We set $G_{i,j}$ for two nearest-neighbors to be present (1) if *both* sites i and j are present (1). Otherwise we set $G_{i,j}$ to zero, that is, if at least one of the sites is empty we set $G_{i,j}$ to be zero. We assume all the sites on the left and right boundaries to be present. This is where current flows in and where the potentials are set. In addition, we assume all the sites on the top and bottom boundaries to be empty. There is therefore no flow in from the top or bottom. We therefore only study percolation from the left to the right.

We have written subroutines to help you with these studies. The function `sitetobond` transforms your percolation matrix z to a bond matrix. The function `FIND_CON` solves the system of equations to find the potentials, V_i, and the

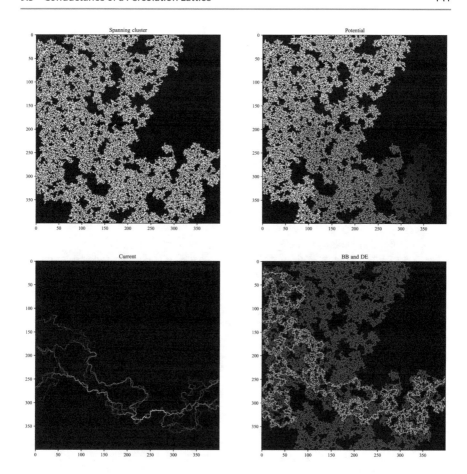

Fig. 9.2 Plots of the spanning cluster, the potential, $V(x, y)$, the absolute value of the current flowing into each site, and the singly connected bonds, the backbone and the dangling ends

function `coltomat` transforms the resulting array of potentials back into a matrix form, $V(x, y)$. The following programs are used to calculate the potentials and currents and visualize the results. The resulting plots are shown in Fig. 9.2.

```
import numpy as np
import matplotlib.pyplot as plt
from scipy.sparse import spdiags, dia_matrix, coo_matrix
from scipy.sparse.linalg import spsolve
from scipy.ndimage import measurements

# Written by Marin Soreng 2004
# Calculates the effective flow conductance Ceff of the
# lattice A as well as the potential V in every site .
def FIND_COND (A , X , Y ):
    V_in = 1.
```

```
        V_out = 0.
        # Calls MK_EQSYSTEM .
        B,C = MK_EQSYSTEM (A , X , Y )
        # Kirchhoff ' s equations solve for V
        V = spsolve(B, C)
        # The pressure at the external sites is added
        # ( Boundary conditions )
        V = np.concatenate((V_in*np.ones(X),V,V_out*np.ones (X)))
        # Calculate Ceff
        # second-last X elements of V multiplied with second-last
        # elem. of A, these are the second last column of the
        # system gives the conductivity of the system per row
        Ceff = np.dot((V[-1-2*X:-1-X]-V_out).T,A[-1-2*X:-1-X,1]) \
          / ( V_in - V_out )
        return V , Ceff


# Sets up Kirchoff ' s equations for the 2 D lattice A .
# A has X * Y rows and 2 columns . The rows indicate the site ,
# the first column the bond perpendicular to the flow direction
# and the second column the bond parallel to the flow direction
#
# The return values are [B , C ] where B * x = C .
# This is solved for the site pressure by x = B \ C .

def MK_EQSYSTEM (A , X , Y ):
    # Total no of internal lattice sites
    sites = X *( Y - 2)
    # Allocate space for the nonzero upper diagonals
    main_diag = np.zeros(sites)
    upper_diag1 = np.zeros(sites - 1)
    upper_diag2 = np.zeros(sites - X)
    # Calculates the nonzero upper diagonals
    main_diag = A[X:X*(Y-1), 0] + A[X:X*(Y-1), 1] + \
      A[0:X*(Y-2), 1] + A[X-1:X*(Y-1)-1, 0]
    upper_diag1 = A [X:X*(Y-1)-1, 0]
    upper_diag2 = A [X:X*(Y-2), 1]
    main_diag[np.where(main_diag == 0)] = 1
    # Constructing B which is symmetric , lower=upper diagonals
    B = dia_matrix ((sites , sites)) # B *u = t
    B = - spdiags ( upper_diag1 , -1 , sites , sites )
    B = B + - spdiags ( upper_diag2 ,-X , sites , sites )
    B = B + B.T + spdiags ( main_diag , 0 , sites , sites )
    # Constructing C
    C = np.zeros(sites)
    #    C = dia_matrix ( (sites , 1) )
    C[0:X] = A[0:X, 1]
    C[-1-X+1:-1] = 0*A [-1 -2*X + 1:-1-X, 1]
    return B , C


def sitetobond ( z ):
    # Function to convert the site network z(L,L) into a
    # (L*L,2) bond network
    # g [i,0] gives bond perpendicular to direction of flow
```

```python
    # g [i,1] gives bond parallel to direction of flow
    # z [ nx , ny ] -> g [ nx * ny , 2]
    nx = np.size (z ,1 - 1)
    ny = np.size (z ,2 - 1)
    N = nx * ny
    gg_r = np.zeros ((nx , ny)) # First , find these
    gg_d = np.zeros ((nx , ny )) # First , find these
    gg_r [:, 0:ny - 1] = z [:, 0:ny - 1] * z [:, 1:ny]
    gg_r [: , ny  - 1] = z [: , ny  - 1]
    gg_d [0:nx - 1, :] = z [0:nx - 1, :] * z [1:nx, :]
    gg_d [nx - 1, :] = 0
    # Then , concatenate gg onto g
    g = np.zeros ((nx *ny ,2))
    g [:, 0] = gg_d.reshape (-1,order='F').T
    g [:, 1] = gg_r.reshape (-1,order='F').T
    return g

def coltomat (z, x, y):
    # Convert z(x*y) into a matrix of z(x,y)
    # Transform this onto a nx x ny lattice
    g = np.zeros ((x , y))
    for iy in range(1,y):
        i = (iy - 1) * x + 1
        ii = i + x - 1
        g[: , iy - 1] = z[ i - 1 : ii]
    return g


# Generate spanning cluster (l - r spanning )
lx = 400
ly = 400
p = 0.5927
ncount = 0
perc = []

while (len(perc)==0):
    ncount = ncount + 1
    if (ncount >100):
        break
    z=np.random.rand(lx,ly)<p
    lw,num = measurements.label(z)
    perc_x = np.intersect1d(lw[0,:],lw[-1,:])
    perc = perc_x[np.where(perc_x > 0)]
    print("Percolation attempt", ncount)
zz = np.asarray((lw == perc[0]))
# zz now contains the spanning cluster
zzz = zz.T # Transpose
g = sitetobond ( zzz ) # Generate bond lattice
V, c_eff = FIND_COND (g, lx, ly) # Find conductivity
x = coltomat ( V , lx , ly ) # Transform to nx x ny lattice
V = x * zzz
g1 = g[:,0]
g2 = g[: ,1]
z1 = coltomat( g1 , lx , ly )
```

```
z2 = coltomat ( g2 , lx , ly )

# Plot results
plt.figure(figsize=(16,16))
ax = plt.subplot(2,2,1)
plt.imshow(zzz, interpolation='nearest')
plt.title("Spanning cluster")
plt.subplot(2,2,2, sharex=ax, sharey=ax)
plt.imshow(V, interpolation='nearest')
plt.title("Potential")

# Calculate current from top to down from the potential
f2 = np.zeros ( (lx , ly ))
for iy in range(ly -1):
    f2[: , iy ] = ( V [: , iy ] - V [: , iy +1]) * z2 [: , iy ]
# Calculate current from left to right from the potential
f1 = np.zeros ( (lx , ly ))
for ix in range(lx-1):
    f1[ ix ,:] = ( V [ ix ,:] - V [ ix +1 ,:]) * z1 [ ix ,:]
# Find the sum of (absolute) currents in and out of each site
fn = np.zeros (( lx , ly ))
fn = fn + abs ( f1 )
fn = fn + abs ( f2 )
# Add for each column (expt leftmost) the offset up-down current
fn [: ,1: ly ] = fn [: ,1: ly ] + abs ( f2 [: ,0: ly -1])
# For the left-most one, add the inverse potential
#   multiplied with the spanning cluster bool information
fn [: ,0] = fn [: ,0] + abs (( V [: ,0] - 1.0)*( zzz [: ,0]))
# For each row (expt topmost) add the offset left-right current
fn [1: lx ,:] = fn [1: lx ,:] + abs ( f1 [0: lx -1 ,:])
# Plot results
plt.subplot(2,2,3, sharex=ax, sharey=ax)
plt.imshow(fn, interpolation='nearest')
plt.title (" Current ")
# Singly connected
zsc = fn > (fn.max() - 1e-6)
# Backbone
zbb = fn>1e-6
# Combine visualizations
ztt = ( zzz*1.0 + zsc*2.0 + zbb*3.0 )
zbb = zbb / zbb.max()
plt.subplot(2,2,4, sharex=ax, sharey=ax)
plt.imshow(ztt, interpolation='nearest')
plt.title (" SC, BB and DE ")
```

Measuring the Conductance

We can now use this program to measure the conductance $G(p, L)$ of the system
and how it varies with both p and L. The idea is to calculate G from $G = I/V$,
where we select V and calculate I using the program. We find I as the sum of

all the currents escaping (or entering) the system. In the program, we have set the potential on the left side to be 1. We recall that we describe positions with the index $j = x + yL$, where the left side corresponds to $x = 0$ and therefore $j = yL$, which we write at iL with $i = y$ is the position along the y-axis. The potentials along the left side are therefore $V_{iL} = 1$ for $i = 0, 1, \ldots, L - 1$. The conductance from site iL into a site on the right, that is, into $iL + 1$, is $G_i L, iL + 1$, which is 1. The current into the system from the left side, that is from site iL into site $iL + 1$, is $I_{iL,iL+1} = G_{iL,iL+1}(V_{iL} - V_{iL+1})$. The total current I into the system is therefore:

$$I = \sum_{i=0}^{L-1} I_{iL,iL+1} = \sum_{i=0}^{L-1} G_{iL,iL+1} (V_{iL} - V_{iL+1}) = \sum_{i=0}^{L-1} (V_{iL} - V_{iL+1}) \ .$$

(9.14)

We use the following program to find the conductance, $G(p, L)$, for an $L \times L$ system for $L = 400$, as well as the density of the spanning cluster $P(p, L)$.

```python
import numpy as np
import matplotlib.pyplot as plt
from scipy.ndimage import measurements
Lvals = [400]
pVals = np.logspace(np.log10(0.58), np.log10(0.85), 20)
C = np.zeros((len(pVals),len(Lvals)),float)
P = np.zeros((len(pVals),len(Lvals)),float)
nSamples = 600
G = np.zeros(len(Lvals))
for iL in range(len(Lvals)):
    L = Lvals[iL]
    lx = L
    ly = L
    for pIndex in range(len(pVals)):
        p = pVals[pIndex]
        ncount = 0
        for j in range(nSamples):
            ncount = 0
            perc = []
            while (len(perc)==0):
                ncount = ncount + 1
                if (ncount > 1000):
                    print("Couldn't make percolation cluster")
                    break
                z=np.random.rand(lx,ly)<p
                lw,num = measurements.label(z)
                perc_x = np.intersect1d(lw[0,:],lw[-1,:])
                perc = perc_x[np.where(perc_x > 0)]
            if len(perc) > 0: # Found spanning cluster
                area = measurements.sum(z, lw, perc[0])
                P[pIndex,iL] = P[pIndex,iL] + area # Find P(p,L)
                zz = np.asarray((lw == perc[0])) # zz=spanning
                zzz = zz.T
```

```
              g = sitetobond (zzz) # Generate bond lattice
              Pvec, c_eff = FIND_COND(g, lx, ly)
              C[pIndex,iL] = C[pIndex,iL] + c_eff
        C[pIndex,iL] = C[pIndex,iL]/nSamples
        P[pIndex,iL] = P[pIndex,iL]/(nSamples*L*L)
plt.plot(pVals,C[:,-1],'-ob',label='$C$')
plt.plot(pVals,P[:,-1],'-or',label='$P$')
plt.legend()
plt.xlabel(r"$p$")
plt.ylabel(r"$g,P$")
```

The resulting behavior for $L = 400$ and $M = 600$ different realizations is shown in Fig. 9.3. We observe two things from this plot: First we see that the behaviors of $G(p, L)$ and $P(p, L)$ are qualitatively different around $p = p_c$: $P(p, L)$ increases very rapidly as $(p - p_c)^\beta$ where β is less than 1. However, it appears that $G(p, L)$ increases more slowly. Indeed, from the plot it looks as if $G(p, L)$ increases as $(p - p_c)^x$ with an exponent x that is larger than 1. How can this be? Why does the density of the spanning cluster increase very rapidly, but the conductance increases much slower? This may be surprising, but we will develop an explanation for this observation in the following.

Conductance and the Density of the Spanning Cluster

For an infinite system, that is when $L \rightarrow \infty$, we cannot define a conductance G. Instead, we must describe the system by its conductivity $g = L^{d-2}G$ (see (9.1)). The two-dimensional system is a special case where the conductance and the

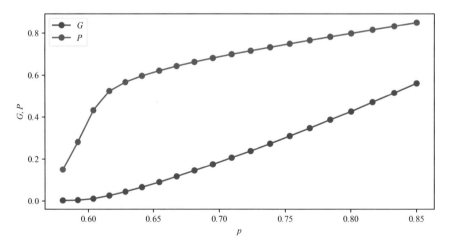

Fig. 9.3 Plots of the conductance $G(p, L)$ and the density of the spanning cluster $P(p, L)$ for $L = 400$

conductivity are identical. However, in general, we need to use this transformation to relate G and g.

For an infinite system, we know that for $p < p_c$ there will be no spanning cluster. The effective conductivity is therefore zero. When p is close to 1, the density of the spanning cluster will be proportional to p, and we also expect the conductance to be proportional to p in this range. This may lead us to assume that the density of the spanning cluster and the conductance of the sample are proportional also when p is close to p_c. However, our direct measurements above (originally done by Last and Thouless [24]) show that P and G are not proportional when p approaches p_c.

We have the tools to understand this behavior. The spanning cluster consists of the backbone and dangling ends. However, it is only the backbone that contributes to conductance of the sample. We could remove all the dangling ends, and still get the same behavior the conductance. This suggests, that it is the scaling behavior of the backbone that is important for the conductance. However, we have found that the mass-scaling exponent of the backbone, D_B, is smaller than D, the mass scaling exponent for the spanning cluster. This indicates that most of the mass of the spanning cluster is found in the dangling ends. This is the reason for the difference between the behavior of $P(p)$, and $G(p)$ for p close to p_c. In the following we will develop a detailed scaling argument for the behavior of the conductance G and the conductivity g of the percolation system.

9.4   Scaling Arguments for Conductance and Conductivity

We will now use the same scaling techniques we introduced to find the behavior of $P(p, L)$ to develop a theory for the conductance $G(p, L)$. First, we realize that we instead of using p as the variable, we may describe the conductance as a function of ξ and L: $G(p, L) = G(\xi, L)$. Second, we realize that the system can only be conducting when there is a spanning cluster, that is, for $p \geq p_c$. We will address two limiting behaviors: (i) the case $L \gg \xi$ and (ii) the case $\xi \gg L$, which means that p is close to p_c.

Scaling Argument for $p > p_c$ and $L \gg \xi$

When $L \gg \xi$ the system is effectively homogeneous over length scales larger than ξ. , we know that over length scales larger than ξ, the system is effectively homogeneous. If were subdivide the system into cells of size ξ, we get a total of $(L/\xi)^d$ effectively homogeneous cells.

For a homogeneous system of ℓ^d boxes of size ℓ, the conductance is $G = \ell^{d-2}G_\ell$, where G_ℓ is the conductance of a single box. We apply the same principle to this system: The conductance $G(\xi, L)$ is given as

$$G(\xi, L) = (\frac{L}{\xi})^{d-2}G(\xi, \xi) \,, \tag{9.15}$$

where $(L/\xi) = \ell$ is the number of boxes and $G(\xi, \xi) = G_\ell$ is the conductance of a single box. We recognize $G(\xi, \xi)$ as the conductance of a system where the correlation length ξ equals the system size, that is, $L = \xi$. We can then find the conductivity $g(\xi, L)$ from (9.1):

$$g(\xi, L) = L^{-(d-2)}G(\xi, L) = \frac{G(\xi, \xi)}{\xi^{d-2}} \ . \tag{9.16}$$

What is $G(\xi, \xi)$? A system with correlation length equal to the system size is indistinguishable from a system at $p = p_c$. The conductance $G(\xi, \xi)$ is therefore the conductance of the spanning cluster at $p = p_c$ in a system of size $L = \xi$. Let us therefore find the conductance of a finite system of size L at the percolation threshold.

Conductance of the Spanning Cluster

What is the conductance, $G(\infty, L)$, of the spanning cluster at $p = p_c$? We know that the spanning cluster consists of the backbone and the dangling ends, and that only the backbone will contribute to the conductivity. The backbone can be described by the blob model (see Sect. 8.2 for a discussion of the blob model): The backbone consists of blobs of bonds in parallel, and links of singly connected bonds between them.

Starting from a scaling hypothesis for the conductance, we will derive the consequences of this assumption, and then test these consequences to see if the data is consistent with our assumption. This will then corroborate the hypothesis. Our scaling hypothesis will be that the conductance of a system of size L at p_c can be described by a scaling exponent $\tilde{\zeta}_R$:

$$G(\infty, L) \propto L^{-\tilde{\zeta}_R} \ . \tag{9.17}$$

Finding Bounds for the Scaling Behavior In many cases, we cannot find the scaling exponents directly, but we may be able to find bounds for the scaling exponents. We will pursue this approach here. We will find bounds for the scaling of $G(\infty, L)$, and use them to determine bounds for the exponent $\tilde{\zeta}_R$.

Lower Bound for the Scaling Exponent First, we know that the spanning cluster consists of blobs in series with the singly connected bonds. This implies that the resistivity $R = 1/G$ of the spanning cluster is given as the resistivity of the singly connected bonds R_{SC} plus the resistivity of the blobs, R_{blob} since resistances are added for a series of resistances:

$$1/G = R = R_{SC} + R_{blob} \ . \tag{9.18}$$

This implies that $R > R_{SC}$. The singly connected bonds are connected in series, one after another. Their total resistance is the sum of the resistances of each bond, which is the resistance of a single bond, multiplied with the number of sites, M_{SC}. Because R_{blob} is positive, we see from (9.18) that

$$R_{SC} = M_{SC} = R - R_{blob} < R \ . \tag{9.19}$$

Because $M_{SC} \propto L^{D_{SC}}$, and we have assumed that $R \propto L^{\tilde{\zeta}_R}$, we find that

$$L^{D_{SC}} < L^{\tilde{\zeta}_R} \quad \Rightarrow \quad D_{SC} \leq \tilde{\zeta}_R \ . \tag{9.20}$$

This gives a lower bound for the exponent!

Upper Bound for the Scaling Exponent We can find an upper bound by examining the minimal path. The resistance of the spanning cluster will be smaller than or equal to the resistance of the minimal path, since the spanning cluster will have some regions, the blobs, where there are bonds in parallel. Adding parallel bonds will always lower the resistance. Hence, the resistance is smaller than or equal to the resistance of the minimal path. Since the minimal path is a series of resistances in series, the total resistance of the minimal path is the mass of the minimal path multiplied by the resistance of a single bond. Consequently, the resistance of the spanning cluster is smaller than the mass of the minimal path, $M_{\min}$, which we know scales with system size, $M_{\min} \propto L^{D_{\min}}$. We have therefore found an upper bound for the exponent

$$L^{\tilde{\zeta}_R} \propto R \leq L_{min} \propto L^{D_{min}} \ , \tag{9.21}$$

and therefore

$$\tilde{\zeta}_R \leq D_{min} \ . \tag{9.22}$$

Upper and Lower Bound Demonstrate the Scaling Relation We have therefore demonstrated (or proved) the scaling relation

$$D_{SC} \leq \tilde{\zeta}_R \leq D_{min} \ . \tag{9.23}$$

Because this scaling relation shows that the scaling of R is bounded by two power-laws in L, we have also proved that the resistance R is a power-law, and that the exponents are within the given bounds. We notice that when the dimensionality of the system is high, the probability of loops will be low, and blobs will be unlikely. In this case

$$D_{SC} = \tilde{\zeta}_R = D_{min} = D_{max} \ . \tag{9.24}$$

Conductivity for $p > p_c$

By scaling arguments, we have established that the conductance $G(\infty, L)$ of the spanning cluster in a system of size L is described by the exponent $\tilde{\zeta}_R$:

$$G(\infty, L) \propto L^{-\tilde{\zeta}_R} \text{ when } L \leq \xi . \tag{9.25}$$

We use this to find an expression for $G(\xi, \xi)$, which is the conductance of the spanning cluster at $p = p_c$ in a system of size $L = \xi$, by inserting $L = \xi$ in (9.25):

$$G(\xi, \xi) \propto \xi^{-\tilde{\zeta}_R} . \tag{9.26}$$

We insert this in (9.16) in order to establish the behavior of the conductivity, g, for $p > p_c$, finding that

$$g = \frac{G(\xi, \xi)}{\xi^{d-2}} \propto \xi^{-(d-2+\tilde{\zeta}_R)} \tag{9.27}$$

$$\propto (p - p_c)^{\nu(d-2+\tilde{\zeta}_R)} \propto (p - p_c)^{\mu} \tag{9.28}$$

Where we have introduced the exponent μ:

$$\mu = \nu(d - 2 + \tilde{\zeta}_R) . \tag{9.29}$$

We notice that for two-dimensional percolation, any value of $\tilde{\zeta}_R$ larger than $1/\nu$ will lead to $\mu > 1$, which was what was observed in Fig. 9.3. The exponent μ is therefore larger than 1, which is significantly different from the exponent β, which is less than 1, which describes the mass of the spanning cluster.

Can the Results Be Generalized? We have therefore explained the difference between how $P(p, L)$ and $G(p, L)$ (or $g(p, L)$) scales with $(p - p_c)$ close to p_c. This is a useful insight and a useful result that provides important information about how a random porous material behaves just as flow is starting to occur through it. Notice that when we study percolation systems, we have generally assumed that the porosity is uncorrelated. For real systems, the porosity may have correlations due to the physical processes that have generated the porosity or the underlying materials. However, when we know how to describe uncorrelated systems like the percolation system, we may use similar theoretical, scaling and computational approaches to study the behavior of real and possibly correlated systems.

9.5   Renormalization Calculation

Another theoretical approach to address and understand the behavior of the system is through the renormalization calculation. Here, we will use the renormalization approach for a square bond lattice in order to estimate the exponent $\tilde{\zeta}_R$.

In order to apply the renormalization approach, we calculate the average resistance $\langle R \rangle$ of a 2×2 cell. We use the H-cell approach and only look at percolation in the horizontal direction. The various configurations c and their degeneracy $g(c)$ is illustrated in Table 9.1. (The degeneracy is the number of configurations in the same class). We assume that the resistance of a single bond is R_0. The average resistance, $\langle R \rangle$, of the renormalized cell is then the probability of the renormalized cell to be occupied, p', multiplied with the resistance of the renormalized cell, R', so that $p'R' = \langle R \rangle$. Using the scaling relation for the resistance, $R \propto L^{\tilde{\zeta}_R}$, we can determine the exponent from

$$\tilde{\zeta}_R = \frac{\ln R'}{\ln b} \ . \tag{9.30}$$

where all the values are calculated for p^*, which we recall is $p^* = 1/2$ for this scheme. The renormalization scheme and the values used are shown in Table 9.1, where we use $p' = p^* = 1/2$ to calculate R'. The resulting value for the renormalized resistance is

$$R' = \frac{1}{p'} \sum_c g(c) P(c) R(c) \tag{9.31}$$

$$= \frac{1}{p'} \left(\frac{1}{2} \right)^5 \left(1 + 1 + 4 \cdot \frac{5}{3} + 2 \cdot 2 + 2 \cdot 3 + 4 \cdot 2 + 2 \cdot 2 \right) \tag{9.32}$$

$$\simeq 1.917 \ . \tag{9.33}$$

Table 9.1 Renormalization scheme for the scaling of the resistance R in a random resistor network. The value $R(c)$ gives the resistance of configuration c, and $g(c)$ is the degeneracy, that is, the number of such configurations

c		$P(c)$	$g(c)$	R_c
1		$p^5(1-p)^0$	1	1
2		$p^4(1-p)^1$	1	1
3		$p^4(1-p)^1$	4	5/3
4		$p^3(1-p)^2$	2	2
5		$p^3(1-p)^2$	2	3
6		$p^3(1-p)^2$	4	2
7		$p^2(1-p)^3$	2	2

Consequently, the exponent $\tilde{\zeta}_R$ is given by

$$\tilde{\zeta}_R \simeq \frac{\ln 1.917}{\ln 2} \simeq 0.939 \; . \tag{9.34}$$

This value is consistent with the scaling bounds set by the scaling relation in (9.24).

9.6   Finite Size Scaling

In general, the conductance and the conductivity is related by:

$$G(p, L) = L^{d-2} g(p, L) \tag{9.35}$$

We found that the scaling of the conductivity is:

$$g \propto (p - p_c)^\mu \propto \xi^{-\mu/\nu} \; , \tag{9.36}$$

with the exponent μ given as $\mu = \nu \left(d - 2 + \tilde{\zeta}_R \right)$.

How can we use this scaling behavior as a basis for a finite-size scaling ansatz? We extend the behavior of the infinite system to the finite size system by the introduction of a finite size scaling function $f(L/\xi)$:

$$g(\xi, L) = \xi^{-\mu/\nu} f\left(\frac{L}{\xi}\right) \; . \tag{9.37}$$

We find the behavior of the scaling function, $f(u)$, by addressing the limiting cases. When $\xi \to \infty$, we know that $g(\xi, L)$ will only depend on L, which means that $f(L/\xi)$ must cancel the $\xi^{-\mu/\nu}$ term, that is, $f(u) \propto u^{-\mu/\nu}$ for $u \ll 1$. Similarly, when $\xi \ll L$, we know that $g(\xi, L)$ will only depend on ξ, which means that $f(L/\xi)$ must be a constant. The scaling function $f(u)$ therefore has the form

$$f(u) = \begin{cases} \text{const.} & \text{when } u \gg 1, \text{ that is } L \to \infty \\ u^{-\mu/\nu} & \text{when } u \ll 1, \text{ that is } \xi \to \infty \end{cases} \tag{9.38}$$

Finite-Size Scaling Observations

How does the scaling ansatz correspond to the observations? We can use the program we have developed to measure the conductivity as a function of both

p and system size L. The following program has been modified for this type of measurement:

```python
import numpy as np
import matplotlib.pyplot as plt
from scipy.ndimage import measurements
from matplotlib.colors import ListedColormap
Lvals = [25,50,100,200,400]
pVals = np.logspace(np.log10(0.58), np.log10(0.85), 20)
C = np.zeros((len(pVals),len(Lvals)),float)
P = np.zeros((len(pVals),len(Lvals)),float)
nSamples = 600
mu = np.zeros(len(Lvals))
for iL in range(len(Lvals)):
    L = Lvals[iL]
    for pIndex in range(len(pVals)):
        p = pVals[pIndex]
        ncount = 0
        for j in range(nSamples):
            ncount = 0
            perc = []
            while (len(perc)==0):
                ncount = ncount + 1
                if (ncount > 1000):
                    print("Couldn't make percolation cluster")
                    break
                z=np.random.rand(L,L)<p
                lw,num = measurements.label(z)
                perc_x = np.intersect1d(lw[0,:],lw[-1,:])
                perc = perc_x[np.where(perc_x > 0)]
            if len(perc) > 0:
                zz = np.asarray((lw == perc[0]))
                # zz now contains the spanning cluster
                zzz = zz.T
                #    # Generate bond lattice from this
                g = sitetobond ( zzz )
                #    # Generate conductivity matrix
                Pvec, c_eff = FIND_COND(g, L, L)
                C[pIndex,iL] = C[pIndex,iL] + c_eff
        C[pIndex,iL] = C[pIndex,iL]/nSamples
for iL in range(len(Lvals)):
    L = Lvals[iL]
    plt.plot(pVals,C[:,iL],label="L="+str(L))
plt.xlabel(r"$p$")
plt.ylabel(r"$g(p,L)$")
plt.legend()
```

The results for $L = 25, 50, 100, 200, 400$ are shown in Fig. 9.4. Here, we plot both the raw data, $g(p, L)$, and the behavior of $g(p_c, L)$ as a function of L on a $\log - \log$-scale, showing that $g(p_c, L)$ indeed scales as a power-law with L.

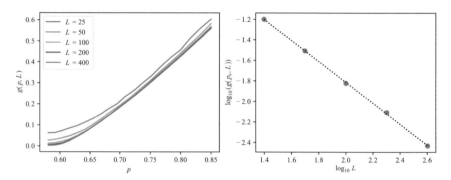

Fig. 9.4 (**a**) Illustration of the conductivity $g(p, L)$ as a function of p for L = 25, 50, 100, 200, 400. (**b**) We see that at p_c the conductivity $g(p_c, L)$ is scaling according to $g \propto L^{-\mu/\nu}$

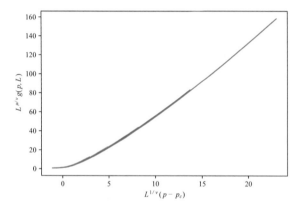

Fig. 9.5 Finite-size data scaling collapse for $g(p, L)$ showing the validity of the scaling ansatz

Scaling Data Collapse   We can also test the scaling ansatz by plotting a finite-size scaling data collapse. We expect that the conductivity will behave as

$$g(p, L) = L^{-\mu/\nu} \tilde{f} (L/\xi) , \qquad (9.39)$$

which we can rewrite by introducing $\xi = \xi_0(p - p_c)^{-\nu}$ to get:

$$g(p, L) = L^{-\mu/\nu} \tilde{f} \left(\left(L^{1/\nu}(p - p_c) \right)^{\nu} \right) . \qquad (9.40)$$

In Fig. 9.5 we demonstrate that this scaling form is valid by getting a data collapse when we plot $L^{\mu/\nu} g(p, L)$ as a function of $L^{1/\nu}(p - p_c)$.

Estimating the Exponent μ from the Data   We can also use the results from the simulations to measure μ directly by plotting $g(p, L)$ as a function of $(p - p_c)$

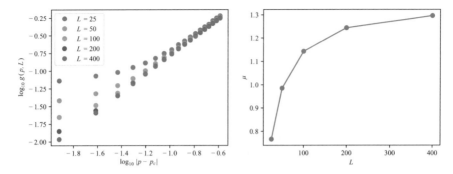

Fig. 9.6 (a) Plot of $g(p, L)$ for increasing values of L. (b) Plot of the exponent μ calculated by a linear fit for increasing system sizes L

and fitting a linear function on a log-log plot. We do this for increasing values of L in Fig. 9.6 (Notice that the curves for small values of L clearly are not linear, and we should, ideally, have fitted the linear curve to only the part of the curve that is approximately linear. We will not address methods to do this here, but you should develop such methods in your own research.).

Implications of the Scaling Ansatz Our conclusion is that the conductivity is a function of p, but also of system size, which implies that the conductivity of a disordered system close to p_c is not a simple material property as we are used to. We therefore need to address the scaling behavior of the system in detail in order to understand the behavior of the conductivity and the conductance of the system.

9.7 Internal Distribution of Currents

When we solve the flow problem on a percolation cluster, we find a set of currents $I_b = I_{i,j}$ for each bond $b = (i, j)$ on the backbone. For all other bonds, the currents will be identically zero. How can we describe the distribution of currents on the backbone?

For electrical flow, the conservation of energy is formulated in the expression:

$$RI^2 = \sum_b r_b I_b^2 , \tag{9.41}$$

where R is the total resistance of the system, I is the total current, r_b is the resistivity of bond b and I_b is the current in bond b. We can therefore rewrite the total resistance R as

$$R = \sum_b r_b \left(\frac{I_b}{I}\right)^2 = \sum_b r_b i_b^2 , \tag{9.42}$$

where we have introduced the fractional current $i_b = I_b/I$. We have written the total resistance as a sum of the square of the fractional currents in each of the bonds.

Distribution of Fractional Currents   The fractional current i_b is assigned to each bond of the backbone. We can describe the fractional currents by the probability distribution for various values of i_b by counting the number of bonds $n(i_b)$ having the fractional current i_b. The total number of bonds is the mass of the backbone:

$$\sum_b 1 = M_B \propto L^{D_B} . \tag{9.43}$$

The distribution of fractional currents is therefore given by $P(i_b) = n(i_b)/M_B$. We characterize the distribution $P(i)$ through the moments of the distribution:

$$\langle i^{2q} \rangle = \frac{1}{M_B} \sum_b i_b^{2q} = \frac{1}{M_B} \int i^{2q} n(i) di . \tag{9.44}$$

There is, unfortunately, no general way to simplify this relation, since we do not know whether the function $n(i)$ has a simple scaling form.

Moments of the Distribution of Currents   However, we can address specific moments of the distribution. We know that the mass of the backbone has a fractal scaling with exponent D_B. This corresponds to the zero'th moment of the distribution. We expect (or hypothesize) that at $p = p_c$, the other moments has a scaling form:

$$\sum_b i_b^{2q} \propto L^{y(q)} . \tag{9.45}$$

What can we say about the scaling exponents $y(q)$ for moment q?

- For $q = 0$, the sum is

$$\sum_b (i_b^2)^0 \propto L^{y(0)} \propto L^{D_B} , \tag{9.46}$$

that is, $y(0) = D_B$.

- For $q \to \infty$, the only terms that will be important in the sum are the terms where $i_b = 1$, because all other terms will be zero. The bonds with $i_b = 1$ are the singly connected bonds: all the current passes through these bonds. Therefore, we have

$$\sum_b (i_b^2)^\infty \propto L^{y(\infty)} \propto M_{SC} \propto L^{D_{SC}} , \tag{9.47}$$

and we find that $y(\infty) = D_{SC}$.

- For $q = 1$, we find from (9.42) that the sum is given as the total resistance of the cluster

$$\sum_b (i_b^2)^1 = R \propto L^{\tilde{\zeta}_R} ,\tag{9.48}$$

which implies that $y(1) = \tilde{\zeta}_R$.

Multifractal Distribution The distribution of fractional currents is an example of a *multi-fractal distribution*. The higher moments of this distribution have a non-trivial scaling relation

$$M_q = \langle i^{2q} \rangle = \frac{\sum_b i_b^2}{M_B} \propto L^{y(q)-D_B} .\tag{9.49}$$

Because each term in the sum $\sum_b (i_b)^{2q}$ is monotonically decreasing in q, the sum is also monotonically decreasing. We can therefore illustrate the curve $y(q)$ as in Fig. 9.7, where we see that $y(q)$ is a non-trivial function of q. This is in contrast to a *unifractal distribution*. We have seen unifractals in e.g. cluster number density. The moments of the cluster number density has the form $M_q \propto \xi^{xq}$ with $x = \gamma(\beta + 1)/\nu$. This means that all the moments are effectively described by a single exponent, x. We call such distributions *unifractal*, whereas distributions where the relationship is non-linear, such as for $y(q)$, we call *multifractal*.

In real resistor-networks, the case is even more complex, because the resistivity is due to impurities, and the impurities diffuse. Therefore, the fluctuations in the resistivity will also have a time-dependent part. This is the origin of thermal noise in the circuit. If we keep the total current I constant, fluctuations in the resistivity will lead to fluctuations in the voltage.

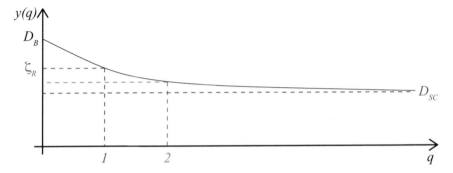

Fig. 9.7 Illustration of the exponents $y(q)$ characterizing the scaling of the moments of the distribution of fractional currents, as a function q, the order of the moment

9.8   Real Conductivity

So far we have addressed conductivity of a percolation cluster. That is a system where the local conductances (or permeabilities) are either zero or a given constant conductance. That is, we have studied a system with local conductances $G_{i,j}$ so that

$$G_b = G_{i,j} = \begin{cases} 1 & \text{with probability } p \\ 0 & \text{with probability } 1 - p \end{cases} . \qquad (9.50)$$

However, in practice, we want to address systems with some distribution of conductances, such as a binary mixture of good and bad conductors, with conductances:

$$G_b = G_{i,j} = \begin{cases} G_2 & \text{with probability } p \\ G_1 & \text{with probability } 1 - p \end{cases} . \qquad (9.51)$$

Superconductor Networks However, in order to address this problem, let us first look at the conjugate problem to the random resistor network, the random superconductor network. We will assume that the conductances are

$$G_b = G_{i,j} = \begin{cases} \infty & \text{with probability } p \\ 1 & \text{with probability } 1 - p \end{cases} . \qquad (9.52)$$

In this case, we expect the conductance to diverge when p approaches p_c from below, and that the conductance is infinite when $p > p_c$. It can be shown that the behavior for the random superconductor network is similar to that of the random resistor network, but that the exponent describing the divergence of the conductance (and consequently conductivity) when p approaches p_c is s:

$$G \propto (p_c - p)^{-s} , \qquad (9.53)$$

Combining the Two Approaches How can we address both these problems? For any system with a finite smallest conductance, $G_<$, we can always use the smaller conductance as the unit for conductance, and write the functional form for the conductance of the whole system as

$$G(G_1, G_2, p) = \left(\frac{G(\frac{G_1}{G_1}, \frac{G_2}{G_1}, p)}{G_1} \right) = G(\frac{G_2}{G_1}, p) , \qquad (9.54)$$

We will make a scaling ansatz for the general behavior of G:

$$G = G_2(p - p_c)^\mu f_\pm \left(\frac{(\frac{G_1}{G_2})}{(p - p_c)^y} \right) , \qquad (9.55)$$

where the exponent y is yet to be determined.

- The *random resistor network* we studied above corresponds to $G_1 \to 0$, and $G_2 = c$. In this case, we retrieve the scaling behavior for p close to p_c, by assuming that $f_+(0)$ is a constant.
- For the *random superconductor network*, the conductances are $G_2 \to \infty$, and $G_1 = \text{const.}$. We will therefore need to construct $f_-(u)$ in such a way that the infinite conductance is canceled from the prefactor. That is, we need $f_-(u) \propto u$. We insert this into (9.55), getting

$$G \propto G_2(p - p_c)^\mu \frac{\frac{G_1}{G_2}}{(p - p_c)^y} \propto G_1 |p - p_c|^{\mu+y} . \tag{9.56}$$

Because we know that the scaling exponent should be $\mu + y = -s$ in this limit, we have determined y: $y = -\mu - s$, where μ and s are determined from the random resistor and random superconductor networks respectively.

Finite G_2 and G_1 When $p \to p_c$ the conductance G should approach a constant number when both G_2 and G_1 are finite. However, $p \to p_c$ corresponds to the argument $x \to +\infty$ in the function $f_\pm(x)$. The only way to ensure that the total conductance is finite, is to require that the two dependencies on $(p - p_c)$ cancel exactly. We achieve this by selecting

$$f_\pm(x) \propto x^{\mu/(\mu+s)} . \tag{9.57}$$

We can insert this relation into (9.55), getting

$$G = G_2 |p - p_c|^\mu \left(\frac{\frac{G_1}{G_2}}{|p - p_c|^{\mu+s}}\right)^{\mu/(\mu+s)} , \tag{9.58}$$

which results in

$$G = G_2 \left(\frac{G_1}{G_2}\right)^{\frac{\mu}{\mu+s}} . \tag{9.59}$$

This expression can again be simplified to

$$G(p = p_c) = G_2^{\frac{s}{\mu+s}} G_1^{\frac{\mu}{\mu+s}} , \tag{9.60}$$

In two dimensions, $\mu = s \simeq 1.3$, and the relation becomes:

$$G \propto (G_1 G_2)^{\frac{1}{2}} , \tag{9.61}$$

Exercises

Exercise 9.1 (Density of the Backbone) The backbone of a spanning cluster is
the union of all self-avoiding walks from one side of the cluster to the opposite.
The backbone corresponds to the sites the contribute to the flow conductivity of the
spanning cluster. The remaining sites are the dangling ends.

We call the mass of the backbone M_B, and the density of the backbone $P_B = M_B/L^d$, where L is the system size, and d the dimensionality of the percolation
system. Here, we will study two-dimensional site percolation.

(a) Argue that the functional form of $P_B(p)$ when $p \to p_c^+$ is

$$P_B(p) = P_0(p - p_c)^x , \qquad (9.62)$$

and find an expression for the exponent x. You can assume that the fractal
dimension of the backbone, D_B, is known.
(b) Assume that the functional form of $P_B(p)$ when $p \to p_c^+$ and $\xi \ll L$ is

$$P_B(p) = P_0(p - p_c)^x , \qquad (9.63)$$

Determine the exponent x by numerical experiment. If needed, you may use that
$\nu = 4/3$.

Exercise 9.2 (Flow on Fractals) Use the example programs from the text to study
fluid flow in a percolation system.

(a) Run the example programs provided in the text to visualize the currents on the
spanning cluster.
(b) Modify the program to find the backbone and the dangling ends of the spanning
cluster.
(c) Use the program to find the singly connected bonds in the spanning cluster.

Exercise 9.3 (Conductivity)

(a) Find the conductivity as a function of $p - p_c$. Determine the exponent $\tilde{\zeta}_R$ by
direct measurement.
(b) Find the conductivity at $p = p_c$ as a function of system size L.

Exercise 9.4 (Current Distribution) Use the example programs from the text to
find the currents I_b in each bonds b on a spanning cluster at $p = p_c$, $p = 0.585$,
and $p = 0.60$.

(a) Find the total current I going through the system.

In the following we will study the normalized currents, $i_b = I_b/I$.

(b) Find the distribution $P(i)$ of the normalized currents.

(c) Measure moments of the distribution.

Exercise 9.5 (Bivariate Porous Media) Rewrite the programs in the text to study a bivariate distribution of conductances. That is, for each site, the conductance is 1 with probability p and $g_0 < 1$ with probability $1 - p$.

(a) Visualize the distribution of currents for $g_0 = 0.1$.

(b) Find the conductivity $g(p)$ for $\sigma_0 = 0.1, 0.01$, and 0.001.

(c) Plot $\sigma(p_c)$ as a function of σ_0.

(d) (Advanced) Can you find a way to rescale the conductivities to produce a data-collapse?

Elastic Properties of Disordered Media

10

There are various physical properties that we may be interested in for a disordered material. In the previous chapter, we studied flow problems in disordered materials using the percolation system as a model disordered material. In this chapter we will address mechanical properties of the disordered material.

We will address the behavior of the disordered material in the limit of fractal scaling. In this limit we expect material properties such as Young's modulus to display a non-trivial dependence on system size. That is, we will expect material properties such as Young's modulus to have an explicit system size dependence. We will use the terminology and techniques already developed to study percolation to address the mechanical behavior of disordered systems such as the coefficients of elasticity [4, 11, 20, 28, 40]

10.1 Rigidity Percolation

What are the elastic properties of a percolation system? First, we need to decide on how to convert a percolation system into an elastic system. We will start by modeling an elastic material as a bond lattice, where each bond represents a local elastic element. The element will in general have resistance to stretching and bending. Systems with only stretching stiffness are termed central force lattices. Here, we will address systems with both stretching and bending stiffness.

Models for Stretching and Bending Stiffness We can formulate the effect of bending and stretching through the elastic energy of the system. The energy will have terms that depend on the elongation of bonds—these will be the terms that are related to stretching resistance. In addition, there will be terms related to the bending of bonds. Here we will introduce the bending terms through the angles between bonds. For any two bonds connected to the same site, there will be an

© The Author(s) 2024
A. Malthe-Sørenssen, *Percolation Theory Using Python*, Lecture Notes
in Physics 1029, https://doi.org/10.1007/978-3-031-59900-2_10

Fig. 10.1 Illustration of the
initial bond lattice (dashed,
gray), and the deformed bond
lattice. Three nodes i, j, k are
illustrated. The angle ϕ_{ijk} is
shown. The displacements $\mathbf{u}_i$
and $\mathbf{u}_j$ are shown respectively
with cyan vectors

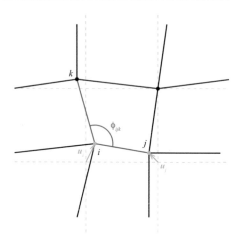

energy associated with changes in the angle of the bond. This can be expressed as

$$U = \sum_{ij} \frac{1}{2}k_{ij}(\mathbf{u}_i - \mathbf{u}_j)^2 + \sum_{ijk} \frac{1}{2}\kappa_{ijk}\phi_{ijk}^2 \, , \tag{10.1}$$

where U is the total energy, the sums are over all particle pairs ij or all particle
triplets ijk. The force constant is $k_{ij} = k$ for bonds in contact and zero otherwise,
and $\kappa_{ijk} = \kappa$ for triplets with a common vertice, and zero otherwise. The vector
$\mathbf{u}_i$ gives the displacement of node i from its equilibrium position. The various
quantities are illustrated in Fig. 10.1

Elastic Modulus Let us address the effective elastic behavior of the percolation
system using a material property such as Young's modulus, E, or the shear modulus,
G. Let us consider a three-dimensional sample with cross-sectional area $A = L^2$ and
length L. Young's modulus, E, relates the tensile stress, σ_{zz}, applied normal to the
surface with area A to the elongation ΔL in the z-direction.

$$\sigma_{zz} = \frac{F_z}{A} = E\frac{\Delta L_z}{L} \, , \tag{10.2}$$

We can therefore write the relation between the force F_z and the elongation ΔL_z as

$$F_z = \frac{EA}{L}\Delta L = \frac{EL^2}{L}\Delta L = L^{d-2}E\Delta L \, . \tag{10.3}$$

We recognize this as a result similar to the relation between the conductance and
the conductivity of the sample, and we will call $K = L^{d-2}E$ the compliance of the
system. We recognize this as being similar to the spring constant of a spring.

Elastic Properties When $p < p_c$ What happens to the compliance of the system as a function of p? When $p < p_c$ there are no connecting paths from one side to another, and the compliance will therefore be zero. It requires zero force F_z to generate an elongation ΔL_z in the system. Notice that we are only interested in the infinitesimal effect of deformation. If we compress the sample, we will of course eventually generate a contacting path, but we are only interested in the initial response of the system.

Elastic Properties When $p > p_c$ When $p \geq p_c$ there will be at least one path connecting the two edges. For a system with a bending stiffness, there will be a load-bearing path through the system, and the deformation ΔL_z of the system requires a finite force, F_z. The compliance K will therefore be larger than zero. We have therefore established that for a system with bending stiffness, the percolation threshold for rigidity coincides with the percolation threshold for connectivity. For a central force lattice, we know that the spanning cluster at p_c will contain may singly connected bonds. These bonds will be free to rotate, and as a result a central force network will have a rigidity percolation threshold which is higher than the connectivity threshold. Indeed, rigidity percolation for central force lattices will have very high percolation thresholds in three dimensions and higher. Here, we will only focus on lattices with bond bending terms.

Behavior of E Close to p_c   Based on our experience with percolation systems, we may hypothesize that Young's modulus will follow a power-law in $(p - p_c)$ when p approaches p_c:

$$E \propto \begin{cases} 0 & \text{for } p < p_c \\ (p - p_c)^\tau & \text{for } p > p_c \end{cases} . \tag{10.4}$$

where τ is an exponent describing the elastic system. We will now use our knowledge of percolation to show that this behavior is indeed expected, and to determine the value of the exponent τ.

Developing a Theory for $E(p, L)$

Let us address the Young's modulus $E(p, L)$ of a percolation system with occupation probability p and a system size L. We could also write E as a function of the correlation length $\xi = \xi(p)$, so that $E = E(\xi, L)$. Young's modulus is in general related to the compliance through $E(\xi, L) = K(\xi, L)L^{d-2}$. We can therefore address the compliance of the system and then calculate Young's modulus.

Dividing the System into Boxes of Size ξ   We will follow an approach similar to what we used to derive the behavior of $P(p, L)$. First, we address the case when the correlation length $\xi \ll L$. In this case, we can subdivide the L^d system into boxes

Fig. 10.2 Illustration of subdivision of a system with $p = 0.60$ into regions with a size corresponding to the correlation length, ξ. The behavior inside each box is as for a system at $p = p_c$, whereas the behavior of the overall system is that of a homogeneous system of boxes of linear size ξ

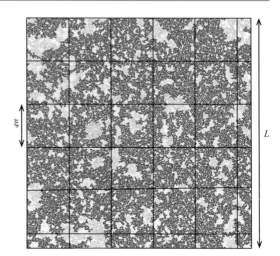

of linear size ξ as illustrated in Fig. 10.2. There will be $(L/\xi)^d$ such boxes. On this scale the system is homogeneous. Each box will have a compliance $K(\xi, \xi)$, and the total compliance will be $K(\xi, L)$.

Compliance of the Combined System We know that the total compliance of n elements in series is $1/n$ times the compliance of a single element. You can easily convince yourself of this addition rule for spring constants, by addressing two springs in series. Similarly, we know that adding n elements in parallel will make the total system n times stiffer, that is, the compliance will be n times the compliance of an individual element. The total compliance $K(\xi, L)$ of this system of $(L/\xi)^d$ boxes is therefore:

$$K(\xi, L) = K(\xi, \xi)(\frac{L}{\xi})^{d-2} . \qquad (10.5)$$

Young's modulus can then be found as

$$E(\xi, L) = L^{-(d-2)} K(\xi, L) = \frac{K(\xi, \xi)}{\xi^{d-2}} . \qquad (10.6)$$

In order to progress further we need to find the compliance $K(\xi, \xi)$. This is the compliance of the percolation system at $p = p_c$ when the system size L is equal to the correlation length ξ. We are therefore left with the problem of finding the compliance of the spanning cluster at $p = p_c$ as a function of system size L.

Compliance of the Spanning Cluster at $p = p_c$

Again, we expect from experience that the compliance will scale with the system size with a dimension $\tilde{\zeta}_K$:

$$K \propto L^{\tilde{\zeta}_K} . \tag{10.7}$$

We will follow our now standard approach: We assume a scaling behavior, establish a set of bounds for K, which will also serve as a proof of the scaling behavior of K, and then use this result to develop a general theory for $K(p, L)$.

Energy, Force and Elongation of the System We will use arguments based on the total energy of the system. The total energy of a system subjected to a force $F = F_z$ resulting in an elongation ΔL is:

$$U = \frac{1}{2} K (\Delta L)^2 , \tag{10.8}$$

where the elongation ΔL is related to the force F through, $\Delta L = F/K$. Consequently,

$$U = \frac{1}{2} K (\frac{F}{K})^2 = \frac{1}{2} \frac{F^2}{K} . \tag{10.9}$$

We can therefore relate the elastic energy of a system subjected to the force F directly to the compliance of that system.

Upper Bound for the Compliance Our arguments will be based on the geometrical picture we have of the spanning cluster when $p = p_c$. The cluster consists of singly connected bonds, blobs, and dangling ends. The dangling ends do not influence the elastic behavior, and can be ignored in our discussion. It is only the backbone that contribute to the elastic properties of the spanning cluster. We can find an upper bound for the compliance by considering the singly connected bonds. The system consists of blobs and singly connected bonds in series. The compliance must include the effect of all the singly connected bonds in series. However, adding the blobs in series as well will only contribute to lowering the compliance. We will therefore get an upper bound on the compliance, by assuming all the blobs to be infinitely stiff, and therefore only include the effects of the singly connected bonds.

Let us therefore study the elastic energy in the singly connected bonds when the cluster is subjected to a force F. The energy, U, can be decomposed in a stretching part, U_s, and a bending part, U_b: $U = U_s + U_b$.

For a singly connected bond from site i to site j, the change in length, $\delta\ell_{ij}$, due to the applied force F is $\delta\ell_{ij} = F/k$, where k is the force constant for a single bond. The energy due to stretching, U_s, is therefore

$$U_s = \sum_{ij} \frac{1}{2} k \delta\ell_{ij}^2 = \sum_{ij} \frac{1}{2} k (\frac{F}{k})^2 = \frac{1}{2} \frac{M_{SC}}{k} F^2 , \tag{10.10}$$

where M_{SC} is the mass of the singly connected bonds.

We can find a similar expression for the bending terms. For a bond between sites i and j, the change in angular orientation, $\delta\phi_{ij}$ is due to the torque $T = r_i F$, where r_i is the distance to bond i in the direction normal to the direction of the applied force F: $\delta\phi_{ij} = T/\kappa$. The contribution from bending to the elastic energy is therefore

$$U_b = \sum_{ij} \frac{1}{2} \kappa (\delta\phi_{ij})^2 = \frac{1}{2} \sum_{ij} \kappa (\frac{r_i F}{\kappa})^2 = \frac{1}{2\kappa} M_{SC} R_{SC}^2 F^2 , \tag{10.11}$$

where

$$R_{SC}^2 = \frac{1}{M_{SC}} \sum_{ij} r_i^2 , \tag{10.12}$$

where the sum is taken over all the singly connected bonds.

The elastic energy of the singly connected bonds is therefore:

$$U_{SC} = (\frac{1}{2k} + \frac{R_{SC}^2}{2\kappa}) M_{SC} F^2 , \tag{10.13}$$

and the compliance of the singly connected bonds is

$$K_{SC} = \frac{F^2}{2U} = \frac{1}{(1/k + R_{SC}^2/\kappa) M_{SC}} . \tag{10.14}$$

which is an upper bound for the compliance K of the system.

Lower Bound for the Compliance We can make a similar argument for a lower bound for the compliance K of the system. The minimal path on the spanning cluster provides the minimal compliance. The addition of any bonds in parallel will only make the system stiffer, and therefore increase the compliance. We can determine the compliance of the minimal path by calculating the elastic energy of the minimal path. We can make an identical argument to what we did above, but we need to replace M_{SC} with the mass, M_{min}, of the minimal path, and the radius of gyration R_{SC}^2 with the radius of gyration of the bonds on the minimal path R_{min}^2.

Kantor [20] has provided numerical evidence that both R_{min}^2 and R_{SC}^2 are proportional to ξ^2. When we are studying the spanning cluster at $p = p_c$

this corresponds to R_{min} and R_{SC} being proportional to L. This shows that the dominating term for the energy is the bending and not the stretching energy when p is approaching p_c.

Bounded Expression for the Compliance K We have therefore determined the scaling relation

$$K_{min} \le K \le K_{SC} \,, \tag{10.15}$$

where we have found that when $L \gg 1$, $K_{min} \propto L^{-(D_{min}+2)}$ and $K_{SC} \propto L^{-(D_{SC}+2)}$. That is:

$$L^{-(D_{min}+2)} \le K(L) \le L^{-(D_{SC}+2)} \,. \tag{10.16}$$

Because $K(L)$ is bounded by two power-laws in L (for all values of L), we have also demonstrated that $K(L)$ also is a power-law in L with an exponent $\tilde{\zeta}_K$ satisfying the relation

$$-(D_{min}+2) \le \tilde{\zeta}_K \le -(D_{SC}+2) \,. \tag{10.17}$$

Finding Young's Modulus $E(p, L)$

This scaling relation gives us $K(p_c, L)$. We use this expression to find $K(\xi, \xi)$, the compliance of a system of size ξ from (10.6):

$$E(\xi, L) = \frac{K(\xi, \xi)}{\xi^{d-2}} \propto \frac{\xi^{\tilde{\zeta}_K}}{\xi^{d-2}} \propto \xi^{\tilde{\zeta}_K - (d-2)} \,. \tag{10.18}$$

We have therefore found a relation for the scaling exponent τ:

$$E(p, L) = \xi^{-(d-2-\tilde{\zeta}_K)} \propto (p - p_c)^{(d-2-\tilde{\zeta}_K)\nu} \propto (p - p_c)^{\tau} \,. \tag{10.19}$$

The exponent τ is therefore in the range:

$$(d - 2 + D_{SC} + 2)\nu \le \tau \le (d - 2 + D_{min} + 2)\nu \,, \tag{10.20}$$

Bounds on the Exponent τ The resulting bounds on the scaling exponents are:

$$(D_{SC} + 2)\,\nu \le \tau \le (D_{min} + 2)\,\nu \,, \tag{10.21}$$

For two-dimensional percolation the exponents are approximately

$$3.41 \le \tau \le 3.77 \,, \tag{10.22}$$

Similarity Between the Flow and the Elastic Problems We see that the bounds are similar to the bounds we found for the exponent $\tilde{\zeta}_R$. This similarity lead Sahimi [29] and Roux [27] to conjecture that the elastic coefficient E and the conductivity g is related through

$$\frac{E}{g} \propto \xi^{-2} \ . \tag{10.23}$$

and therefore that

$$\tau = \mu + 2\nu = (d + \tilde{\zeta}_R)\nu \ . \tag{10.24}$$

which is well supported by numerical studies.

In the limit of high dimensions, $d \geq 6$, the relation $\tau = \mu + 2\nu = 4$ becomes exact. However, we can use as a rule of thumb that the exponent $\tau \simeq 4$ in all dimensions $d \geq 2$.

Diffusion in Disordered Media 11

In this chapter we will study diffusional transport in disordered media. We can model diffusional transport either by solving the diffusion equation or by studying the time developments of random walks—both approaches produce the same results. We will use the statistical approach and study how random walkers spread with time in free space as well as on percolation clusters. We will introduce a scaling theory for the behavior of this process in both space and time—extending our previous scaling approaches and proving us with new tools and insights. We will do this in several steps, starting with a brief introduction to random walks and diffusion in uniform media, then introduce a computational model for random walks on the percolation cluster, and finally apply our full set of tools to develop scaling theories for the observed behavior [13, 15, 26].

11.1 Diffusion and Random Walks in Homogeneous Media

A typical example of a random walk is the random motion of a small dust particle due to random collisions with air molecules, a process called Browian motion. Random walks are general processes that we often use as physical, theoretical or conceptual models.

A Two-Dimensional Random Walk If a random walker starts at $\mathbf{r} = 0$, its position $\mathbf{r}_n$ after n steps can be written as

$$\mathbf{r}_n = \mathbf{r}_0 + \sum_{i=1}^{n} \mathbf{u}_i \ , \tag{11.1}$$

where $\mathbf{u}_i$ is step i. We will usually assume that the steps $\mathbf{u}_i$ are independent and isotropically distributed.

© The Author(s) 2024
A. Malthe-Sørenssen, *Percolation Theory Using Python*, Lecture Notes
in Physics 1029, https://doi.org/10.1007/978-3-031-59900-2_11

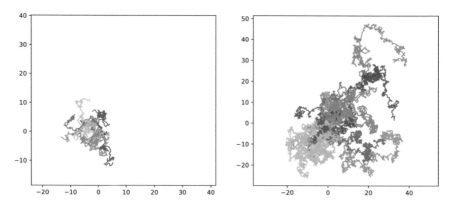

Fig. 11.1 Plots of 10 random walks of size $n = 100$ (left) and $n = 1000$ (right)

Generating a Random Walk We can generate an example of random walk by selecting $\mathbf{u}_i = (x_i, y_i)$, where x_i and y_i are selected from e.g. a uniform random distribution from -1 to 1. The following program calculates and visualizes a random walk starting from the origin. The resulting path is shown in Fig. 11.1. We notice that the random walker spreads out gradually, leaving behind a trace with a complex geometry.

```
import numpy as np
import matplotlib.pyplot as plt
n = 100
u = 2*np.random.rand(n,2)-1
r = np.cumsum(u,axis=0)
plt.plot(r[:,0],r[:,1])
```

Theory for the Time Development of a Random Walk

We can develop a theory for the position $\mathbf{r}_n$ as a function of the number of steps n. For simplicity, we start the walker at the origin, so that $\mathbf{r}_0 = 0$. First, we see find the average position after n steps:

$$\langle \mathbf{r}_n \rangle = \langle \sum_{i=1}^{n} \mathbf{u}_n \rangle = \sum_{i=1}^{n} \langle \mathbf{u}_i \rangle = \mathbf{0} \, , \tag{11.2}$$

where we have used that since $\mathbf{u}_i$ are isotropic, $\langle \mathbf{u}_i \rangle = 0$. This is not surprising, the random walker has the same probability to walk in all directions and therefore does not get anywhere on average.

However, from Fig. 11.1 we see that the extent of the path increases with the number of steps n. We can characterize this using the same measures we used to describe the geometry of the percolation clusters, by measuring r_n^2. We find the average value of r_n^2 using

$$
\begin{aligned}
\langle r_n^2 \rangle &= \langle \mathbf{r}_n \cdot \mathbf{r}_n \rangle \\
&= \langle (\sum_i \mathbf{u}_i) \cdot (\sum_j \mathbf{u}_j) \rangle \\
&= \langle \sum_i \sum_j \mathbf{u}_i \cdot \mathbf{u}_j \rangle \\
&= \langle \sum_{i=j} \mathbf{u}_i \cdot \mathbf{u}_j \rangle + \langle \sum_{i \neq j} \mathbf{u}_i \cdot \mathbf{u}_i \rangle \\
&= \sum_i \langle \mathbf{u}_i \cdot \mathbf{u}_i \rangle + \sum_{i \neq j} \underbrace{\langle \mathbf{u}_i \cdot \mathbf{u}_i \rangle}_{=0} \\
&= n\delta^2 \,,
\end{aligned}
\tag{11.3}
$$

where $\langle \mathbf{u}_i \cdot \mathbf{u}_i \rangle = \delta^2$ is a property of the distribution of $\mathbf{u}_i$ corresponding to the variance of the distribution. And where we have used that because $\mathbf{u}_i$ and $\mathbf{u}_j$ are independent, the average of their product is equal to the product of their averages:

$$
\langle \mathbf{u}_i \cdot \mathbf{u}_j \rangle = \langle \mathbf{u}_i \rangle \cdot \langle \mathbf{u}_j \rangle = \mathbf{0} \cdot \mathbf{0} = 0 \,.
\tag{11.4}
$$

Consequently, we have shown that $r_n^2 = n\delta^2$. This is a *very general result*. We have found that the extent of the diffusion path increases slowly with the number of steps: $r_n = \delta n^{1/2}$. This result is valid in any dimension as long as the two basic assumptions are satisfied: The individual steps are independent and each individual step has an isotropic distribution so that the average displacement from a single step is zero.

The Dimension of the Random Walk Here we have demonstrated that the size of the random walk, measured as r^2, is proportional to the number of elements in the random walk. This is similar to the way we measured the size of a cluster using the radius of gyration of the cluster. Indeed, it can be shown that these two definitions give the same relation $r_n^2 = b^2 n$, where b is a constant of unit length that describes the distribution of a single step. We realize that n is the number of elements in the random walk, corresponding to s, the number of sites in a cluster. We have therefore found that $r_n = bn^{1/2}$, or similarly, that $n = (r_n/b)^2 \propto r^{D_w}$. This implies that the dimension, D_w, of the random walk always is $D_w = 2$, independent of the embedding dimension d. This means that for $d = 1$ the random walk will overfill space. Indeed, we expect it to step on top of itself repeatedly. For $d = 2$ the random walk will just fill space since $D_w = d$, whereas for $d = 3$ and higher dimensions the random walk will fill a diminishing portion of space. Just like the spanning cluster

had a smaller scaling exponent than the spatial dimension, and hence the density of
the spanning cluster decreased for larger systems.

Continuum Description of a Random Walker

We can also describe the motion of the random walker through the probability
density $P(\mathbf{r}, t)$, where $P(\mathbf{r}, t)\,d\mathbf{r}\,dt$ is the probability for the random walker to be in
the volume $\mathbf{r}\,d\mathbf{r}$ in the time period t to $t + dt$.

For a random walker on a grid, the probability to be at a grid position i is given
as $P_i(t)$. The probability for the walker to be at a position i at the time $t = t + \delta t$ is
then

$$P_i(t + \delta t) = P_i(t) + \sum_j [\sigma_{j,i} P_j(t) - \sigma_{i,j} P_i(t)]\delta t , \tag{11.5}$$

where the sum is over all neighbors j of the site i. The term $\sigma_{i,j}$ is the transition
probability from site i to site j. The first term in the sum represents the probability
that the walker during the time period δt walks into site i from site j, and the second
term represents the probability that the walker during the time period δt walks from
site i to one of the neighboring sites j.

When $\delta t \to 0$ this equation approaches a differential equation

$$\frac{\partial P_i}{\partial t} = \sum_j [\sigma_{j,i} P_j(t) - \sigma_{i,j} P_i(t)] . \tag{11.6}$$

If we assume that the transition probability is equal for all the neighbors, so that
$\sigma_{i,j} = 1/Z$, where Z is the number of neighbors, the differential equation simplifies
to

$$\frac{\partial P}{\partial t} = D\nabla^2 P , \tag{11.7}$$

which we recognize as the diffusion equation, where the diffusion constant D is
related to the transition probabilities $\sigma_{i,j}$ and Z.

The general solution to this equation is

$$P(\mathbf{r}, t) = \frac{1}{(2\pi Dt)^{d/2}} e^{-r^2/2Dt} = \frac{1}{(2\pi)^{d/2}|\mathbf{R}|^2} e^{-\frac{1}{2}(\frac{r}{|\mathbf{R}|})^2} , \tag{11.8}$$

where we have introduced $|\mathbf{R}| = \sqrt{Dt}$.

It can be shown that the moments of this distribution are

$$\langle r^k \rangle = A_k R(t)^k \propto t^{k/2} , \tag{11.9}$$

and specifically, that

$$\langle r^2 \rangle = \int P(\mathbf{r}, t) r^2 d\mathbf{r} = R^2(t) = Dt .$$ (11.10)

which displays the same relationship as found above between extent, $\langle r^2 \rangle$, and time, t, where the time is $t = n\delta t$ and δt is the time a single step takes.

11.2 Random Walks on Clusters

We now have the basic tools to understand diffusion in homogeneous media: by studying the position $\mathbf{r}(t)$ of a random walker as a function of the number of steps n or the time $t = n\Delta t$, where Δt is the time for a single step.

How can we use this method to study diffusion on a percolation cluster? We want to address how a particle diffuses on the cluster. That is, we want to study how a random walker moves on the occupied sites in the percolation system. We will assume that the walker only can move onto connected neighbor sites in each step.

There are many different ways we can construct such measurements, and as always, we need to be very precise when we define both the experiment and our set of measures. Our plan is to drop a random walker onto a random site in the percolation system and measure the position $\mathbf{r}(t)$ of the walker as a function of time.

Developing a Program to Study Random Walks on Clusters

In order to *study* the behavior we need to develop a program to generate a random walk on top of a percolation lattice, generate many such paths and collect, analyze and visualize the resulting behavior.

The rules for such a walker would be that we select a position at random and then parachute the walker into this position. We start with a percolation system given by the $L \times L$ matrix cluster, where cluster is True in the points where the sites are present. The initial positions, ix, iy, in the x- and y-direction for the walker are therefore random numbers between 0 and $L - 1$ respectively:

```
ix = np.random.randint(L)
iy = np.random.randint(L)
```

where L is the system size. If this site is empty, the walk stops immediately and its length is zero:

```
if not cluster[ix,iy]:
    return
```

Storing the Trajectory of the Walker We store the trace of the walker in two arrays (we need both to handle periodic boundary conditions later): walker_map which consists of the positions ix,iy of the walker for each step, and displacement, which consists of the positions relative to the initial position of the walker.

Random Selection of Next Step How do we select where the walker can move? The walker is restricted to move to nearest neighbor sites that are present. There are several approaches:

- We may select a direction at random and try to move in this direction. If the walker cannot move in this direction it stays put for this step, and then tries again in the next step. In this case, the walker may have many steps without any motion.
- We may find all the directions the walker can possibly move in, and then select one of these directions at random. In this case the walker will move onto a new site in each step.

Both these methods effectively produce the same behavior. We will select the second method. We therefore need to create a list of the possible directions to move in. In order to make this list, we have a list called directions of possible movement directions:

```
directions = np.zeros((2, 4), dtype=np.int64)
# X-dir: east and west, Y-dir: north and south.
directions[0, 0] = 1
directions[1, 0] = 0
directions[0, 1] = -1
directions[1, 1] = 0
directions[0, 2] = 0
directions[1, 2] = 1
directions[0, 3] = 0
directions[1, 3] = -1
```

For each step, we need to collect all the possible steps into a list called neighbor_arr. This is done by the following loop:

```
neighbor = 0
    for idir in range(directions.shape[1]):
        dr = directions[:,idir]
        iix = ix + dr[0]
        iiy = iy + dr[1]
        if 0<=iix<L and 0<=iiy< L and cluster[iix,iiy]:
            neighbor_arr[neighbor] = idir
            neighbor += 1
```

If this list is empty, that is, if neighbor is zero, there are *no possible places to move*. This means that the walker has landed on a cluster of size $s = 1$. In this case, we stop and return with $n = 1$.

Finally, we select one of the `neighbor` directions at random, move the walker into this site, update `walker_map` and `displacement` and repeat the process.

```
# Select random direction from 0 to neighbor-1
randdir = np.random.randint(neighbor)
dir = neighbor_arr[randdir]
ix += directions[0, dir]
iy += directions[1, dir]
step += 1
walker_map[0, step] = ix
walker_map[1, step] = iy
displacement[:,step]=displacement[:,step-1]+\
  directions[:,dir]
```

Here, `step` corresponds to n, the current step number.

Preparing the Function We put this into a function and use the numba library to speed up simulation times.

```
import numba
import numpy as np

@numba.njit(cache=True)
def percwalk(cluster, max_steps):
    """Function performing a random walk on the spanning cluster
    Parameters
    ----------
    cluster : np.ndarray
        Boolean array with 1's signifying a present site
    max_steps : int
        Maximum number of walker steps to perform.
    Returns
    -------
    walker_map : np.ndarray
        A coordinate map of walk, x in [0] and y in [1]
    displacement : np.ndarray
        A coordinate map relative pos., x in [0] and y in [1]
    num_steps : int
        Number of steps performed.
    """
    walker_map = np.zeros((2, max_steps))
    displacement = np.zeros_like(walker_map)
    directions = np.zeros((2, 4), dtype=np.int64)
    neighbor_arr = np.zeros(4, dtype=np.int64)
    # X-dir: east and west, Y-dir: north and south.
    directions[0, 0] = 1
    directions[1, 0] = 0
    directions[0, 1] = -1
    directions[1, 1] = 0
    directions[0, 2] = 0
    directions[1, 2] = 1
    directions[0, 3] = 0
```

```
directions[1, 3] = -1
# Initial random position
Lx, Ly = cluster.shape
ix = np.random.randint(Lx)
iy = np.random.randint(Ly)
walker_map[0, 0] = ix
walker_map[1, 0] = iy
step = 0
if not cluster[ix, iy]: # Landed outside the cluster
    return walker_map, displacement, step
while step < max_steps-1:
    # Make list of possible moves
    neighbor = 0
    for idir in range(directions.shape[1]):
        dr = directions[:,idir]
        iix = ix + dr[0]
        iiy = iy + dr[1]
        if 0<=iix<Lx and 0<=iiy<Ly and cluster[iix,iiy]:
            neighbor_arr[neighbor] = idir
            neighbor += 1
    if neighbor == 0: # No way out, return
        return walker_map, displacement, step
    # Select random direction from 0 to neighbor-1
    randdir = np.random.randint(neighbor)
    dir = neighbor_arr[randdir]
    ix += directions[0, dir]
    iy += directions[1, dir]
    step += 1
    walker_map[0, step] = ix
    walker_map[1, step] = iy
    displacement[:,step]=displacement[:,step-1]+\
        directions[:,dir]
return walker_map, displacement, step
```

Testing the Function Let us test the newly generated function on a few simplified cases. First, we try it on a system with $p = 1$, that is, on a homogeneous system.

```
import numpy as np
import matplotlib.pyplot as plt
L = 50
p = 1
z = np.random.rand(L,L)<p
plt.imshow(z,origin="lower")
walker_map, displacement, steps = percwalk(z,200)
# walker_map is oriented as row-column (ix, iy)
plt.plot(walker_map[1,:steps],walker_map[0,:steps])
```

Walks from 10 such simulations are shown in Fig. 11.2. This looks reasonable and nice, but we do notice that quite a few of these walks reach the boundaries of the system. We may wonder how this finite system size affects the behavior and statistics of the system.

Fig. 11.2 Trajectories of 10 random walks for a (homogeneous) system with $L = 50$ and $p = 1$

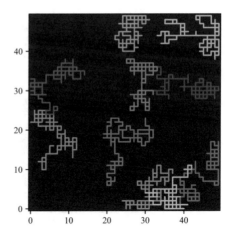

Measuring $r^2(t)$ for a Random Walker The function `percwalk` returns the displacements, $\mathbf{r}_n$, for the walking starting from $\mathbf{r}_0 = 0$. We find r_n^2 and visualize the result for a single walk:

```
L = 50
p = 1
z = np.random.rand(L,L)<p
walker_map, displacement, steps = percwalk(z,200)
r2 = np.sum(displacement**2,axis=0)
t = np.arange(len(r2))
plt.plot(t,r2)
```

The resulting plot is shown in Fig. 11.3. We do not really learn much from this plot. We need to collect more statistics. We need to generate many different walks and then average over all the walks to find a statistically better measure for $r^2(t)$.

Collecting Statistics for $r^2(t)$ We therefore write a small function to generate a given number of clusters with the given p. For each such cluster we will generate a given number of walks. Notice, that we must also specify the maximum number of steps that we model for each walk. The following function implements these features:

```
@numba.njit(cache=True)
def find_displacements(p,L,num_systems,num_walkers,max_steps):
    displacements = np.zeros(max_steps)
    for system in range(num_systems):
        z = np.random.rand(L,L)<p
        for j in range(num_walkers):
            num_steps = 0
            while num_steps <= 1:
                walker_map,displacement,num_steps = \
            percwalk(z,max_steps)
            displacements += np.sum(displacement**2, axis=0)
    displacements = displacements/(num_walkers*num_systems)
    return displacements
```

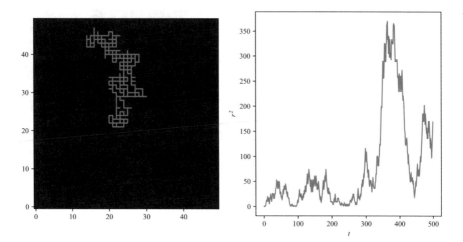

Fig. 11.3 (**a**) Trajectory of a random walk for a (homogeneous) system with $L = 50$ and $p = 1$. (**b**) Plot of the corresponding $r^2(t)$

Notice a few details: If the number of steps is 1 or smaller it means that the walker landed either on an empty size ($n = 0$) or on a single site ($n = 1$). We do not want to include these in our statistics since they provide little information about the behavior of the random walker. We use this program to collect statistics from $M = 500$ random walks of length $n = 10{,}000$ steps on a $L = 100$ system:

```
p = 1.0
L = 100
max_steps = 10000
num_walkers = 500
num_systems = 100
displacements = find_displacements(p,L,num_systems,\
                        num_walkers,max_steps)
dr1 = displacements[1:]
t = np.arange(len(dr1))
plt.loglog(t,dr1)
```

The resulting plot in Fig. 11.4 shows that the system indeed behaves as we expect— for small values of t. However, as t increases, we see that the effect of the finite system size L starts to affect the results. This is because the random walker is limited by the wall and eventually we will be limited the $L \times L$ system. This problem will also arise when we study the percolation system. How can we reduce this problem?

Introducing Periodic Boundary Condition   One way of reducing this problem is by introducing *periodic boundary conditions*. The idea is that is the random walker steps outside the lattice on the left side, it appears on the right-hand side instead. That is, if ix becomes -1, it is instead set to $L - 1$. We implement this in the percwalk function in the following. The resulting plot of $r^2(t)$ in Fig. 11.4 shows

Fig. 11.4 Plot of $r^2(t)$ for a $L = 100$ system with non-periodic and periodic boundary conditions

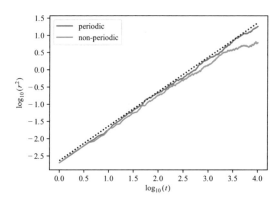

that this solves the problem with the boundaries. This aspect will be even more important when we study percolation systems in non-uniform media.

```python
# With periodic boundary conditions for good statistics
import numba
import numpy as np

@numba.njit(cache=True)
def percwalk(cluster, max_steps):
    """Function performing a random walk on the spanning cluster
    Parameters
    ----------
    cluster : np.ndarray
        Boolean array with 1's signifying a site present
    max_steps : int
        Maximum number of walker steps to perform.
    Returns
    -------
    walker_map : np.ndarray
        A coordinate map of walk, x in [0] and y in [1]
    displacement : np.ndarray
        A coordinate map of relative pos, x in [0] and y in [1]
    num_steps : int
        Number of steps performed.
    """
    walker_map = np.zeros((2, max_steps))
    displacement = np.zeros_like(walker_map)
    directions = np.zeros((2, 4), dtype=np.int64)
    neighbor_arr = np.zeros(4, dtype=np.int64)
    # X-dir: east and west, Y-dir: north and south.
    directions[0, 0] = 1
    directions[1, 0] = 0
    directions[0, 1] = -1
    directions[1, 1] = 0
    directions[0, 2] = 0
    directions[1, 2] = 1
    directions[0, 3] = 0
```

```
    directions[1, 3] = -1
    # Initial random position
    Lx, Ly = cluster.shape
    ix = np.random.randint(Lx)
    iy = np.random.randint(Ly)
    walker_map[0, 0] = ix
    walker_map[1, 0] = iy
    step = 0
    # Check if we landed outside the spanning cluster
    if not cluster[ix, iy]:
        # Return the map with starting pos and nr of steps
        return walker_map, displacement, step
    while step < max_steps-1:
        # Make list of possible moves
        neighbor = 0
        for idir in range(directions.shape[1]):
            dr = directions[:,idir]
            iix = ix + dr[0]
            iiy = iy + dr[1]
            # Periodic BC
            if iix>=Lx:
                iix = iix-Lx
            if iix<0:
                iix = iix+Lx
            if iiy>=Ly:
                iiy = iiy-Ly
            if iiy<0:
                iiy = iiy+Ly
            if cluster[iix, iiy]:
                neighbor_arr[neighbor] = idir
                neighbor += 1
        if neighbor == 0: # No way out, return
            return walker_map, displacement, step
        # Select random direction from 0 to neighbor-1
        randdir = np.random.randint(neighbor)
        dir = neighbor_arr[randdir]
        ix += directions[0, dir]
        iy += directions[1, dir]
        step += 1
        walker_map[0, step] = ix
        walker_map[1, step] = iy
        displacement[:,step]=displacement[:,step-1]+\
                directions[:,dir]
    return walker_map, displacement, step
```

Diffusion on a Finite Cluster for $p < p_c$

We now have all the tools to start studying the behavior of a random walker on top
of a percolation system. We select $p = p_c$ and drop the random walker on a random
position on the lattice. The resulting set of walks from such a simulation can be seen
in Fig. 11.5.

Fig. 11.5 Plot of 30 walks in a $L = 100$ system at $p = p_c$

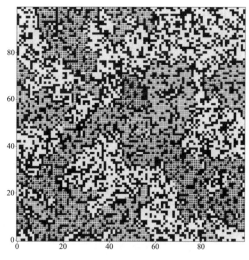

Fig. 11.6 Plots of $r^2(t; p, L)$ for $p = 0.45, 0.50, 0.55, p_c$

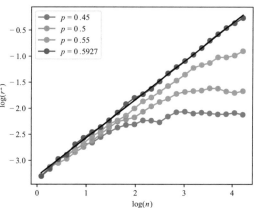

```
L = 100
p = 0.5927
z = np.random.rand(L,L)<p
plt.imshow(z,origin="lower")
for i in range(30):
    walker_map, displacement, steps = percwalk(z,10000)
    plt.plot(walker_map[1,:steps],walker_map[0,:steps],'o')
```

Understanding Behavior for $p > p_c$ We then simulate a larger set of walks for $p = 0.45, 0.50, 0.55, p_c$. The resulting plots of $r^2(t)$ are shown in Fig. 11.6. We see that when $p < p_c$, $r^2(t) \propto t^x$ for some time, but then after some time, $r^2(t)$ crosses over to a constant instead. How can we understand this behavior?

Long-Term Behavior When $p < p_c$ For a single walker that lands on a cluster of size s, we expect that the walker will be limited to walk on this cluster and therefore cannot reach positions that are much further away than R_s. Thus, after a long time, we expect $r^2(t) \propto R_s^2$. If we repeat this experiment many times, each time dropping the walker onto a random occupied point in the system, we need to take the average over all clusters of size s and over all starting positions, to find the average of $r^2(t)$ for all these different walks. If we drop the walker at a random position, the probability for that walker to land on a cluster of size s is $sn(s, p)$, and the contribution from this cluster to $r^2(t)$ after a long time is R_s^2. Therefore, the average $\langle r^2(t) \rangle$ for the walker is:

$$\left[\langle r^2 \rangle \right] \propto \left[R_s^2 \right] = \sum_s R_s^2 sn(s, p) \ . \tag{11.11}$$

We approximate this sum by an integral and replace $n(s, p)$ by the scaling ansatz $n(s, p) = s^{-\tau} F(s/s_\xi)$, getting

$$\left[R_s^2 \right] = \int_1^\infty R_s^2 s s^{-\tau} F(s/s_\xi) ds \ . \tag{11.12}$$

We realize that the function $F(s/s_\xi)$ falls to zero very rapidly when $s > s_\xi$ and it is effectively constant below that, we therefore replace the integral with an integral up to s_ξ:

$$\left[R_s^2 \right] = \int_1^{s_\xi} R_s^2 s s^{-\tau} ds \ . \tag{11.13}$$

We now insert that $R_s^2 \propto s^{2/D}$ and perform the integral, getting:

$$\left[R_s^2 \right] \propto s_\xi^{2/D+2-\tau} \propto s_\xi^{2/D} s_\xi^{2-\tau} \ . \tag{11.14}$$

where we recognize the first factor as $\xi^2 \propto (p - p_c)^{-2\nu}$ and the second factor from (4.33) as $(p - p_c)^\beta$ so that

$$\left[R_s^2 \right] \propto (p - p_c)^{\beta - 2\nu} \ . \tag{11.15}$$

We notice that in this case the average is of R_s^2 over $sn(s, p)$, but when we calculated the correlation length in (5.18) the average was of R_s^2 over $s^2 n(s, p)$, and this is the reason for the appearance of the exponent $\beta - 2\nu$ and not simply -2ν as we got for the correlation length.

Short Term Behavior There is a transition in $r^2(t)$ to $\left[R^2 \right]$ after some crossover time t_0. For times shorter than t_0 we see from Fig. 11.6 that the behavior appears to be that $r^2(t) \propto t^{2k}$ for some exponent $2k$. We notice that as p approaches p_c, the crossover time t_0 increases. All the curves for various p-values appear to have similar, or possibly the same behavior for $t < t_0$.

In Fig. 11.6 we notice that the exponent $2k$ is not 1, as we found for the homogeneous case. It is clearly lower than 1. If we measure it, we find that $2k \simeq 0.66$ and $k \simeq 0.33$. We call this behavior *anomalous diffusion* because the mean squared distance $r^2(t)$ does not grow linearly with time, but with an exponent different than 1. What can we say about the crossover time t_0? We will return to this after examining the case when $p > p_c$.

Diffusion at $p = p_c$

From Fig. 11.6 we also see that for $p = p_c$ the random walk follows $r^2(t) \propto t^{2k}$. This behavior is as expected. For times shorter than t_0, the walker behaves as if it is on p_c, whereas after a long time, $t > t_0$, we start noticing that the walker is restricted when it diffuses on the finite clusters. Another way to think of this is that the crossover time t_0 increases as $p \rightarrow p_c$, and diverges at $p = p_c$. The exponent k is a universal exponent for diffusion on percolation systems. It does not depend on the lattice structure or the rules for connectivity, but it does depend on the embedding dimension d.

Diffusion for $p > p_c$

We can use the same computational approach to study the behavior of the random walker when $p > p_c$. The resulting plots for $p = 0.8, 0.75, 0.70, 0.65$ and p_c are shown in Fig. 11.7. The plots show that when $p > p_c$, for short times the $r^2(t)$ curve follows the behavior for $p = p_c$ with $r^2(t) \propto t^{2k}$ as shown by the solid line. But for a crossover time t_0, the behavior changes and crosses over to a behavior where $r^2(t) \propto t^1$, that is, it crosses over to the behavior of a homogeneous system. How can we understand this?

Fig. 11.7 Plots of $r^2(t; p, L)$ for $p = 0.8, 0.75, 0.70, 0.65, p_c$

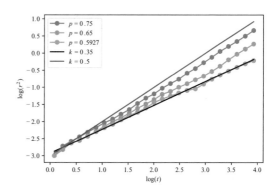

Developing a Model for $p > p_c$ We know that when $p = 1$, the system is homogeneous, and $\langle r^2 \rangle = D(1)t$. We will therefore write the general relation for $p > p_c$:

$$\langle r^2 \rangle = D(p)t \; , r \gg \xi \; . \tag{11.16}$$

What behavior do we expect from $D(p)$? We expect $D(p)$ to increase in a way similar to the density of the backbone or the conductivity g. In fact, the Einstein relation for diffusion relates the diffusion constant to the conductance through:

$$D(p) \propto g(p) \propto (p - p_c)^\mu \; . \tag{11.17}$$

We therefore expect that when $p > p_c$, and the time is larger than a crossover time $t_0(p)$, that the behavior is scaling with exponent μ, identical to that of conductivity. And for a time shorter than the crossover time, the behavior is identical to the behavior at $p = p_c$. We can understand this in the same way as above: When $t < t_c$ the walker does still not experience that the characteristic clusters are limited by a finite characteristic length ξ.

Scaling Theory

Let us develop a scaling theory for the behavior of $\langle r^2 \rangle$. We will assume that when the time is smaller than a cross-over time, the behavior is according to a power-law with exponent $2k$, and that when the time is larger than the cross-over time, the behavior is either that of diffusion with diffusion constant $D(p)$ for $p > p_c$, or it reaches a constant plateau for the case when $p < p_c$.

Let us introduce a scaling ansatz with these properties:

$$\langle r^2 \rangle = t^{2k} f[(p - p_c)t^x] \; . \tag{11.18}$$

Notice that we could have started from any of the end-points, such as from the assumption that

$$\langle r^2 \rangle = (p_c - p)^{\beta - 2\nu} G_1(\frac{t}{t_0}) \; , \tag{11.19}$$

or

$$\langle r^2 \rangle = (p - p_c)^\mu G_2(\frac{t}{t_0}) \; . \tag{11.20}$$

We have two unknown exponents k and x that must be determined from independent knowledge. We will assume that the function $f(u)$ has the behavior

$$f(u) = \begin{cases} \text{const.} & \text{for } |u| \ll 1 \\ u^\mu & \text{for } u \gg 1 \\ (-u)^{\beta-2\nu} & \text{for } u \ll -1 \end{cases} \tag{11.21}$$

Let us now address the various limits in order to determine the scaling exponents k and x in terms of known exponents.

Scaling Behavior in the Limit $p > p_c$ First, we know that when $p > p_c$, that is when $u \gg 1$, we have that

$$\langle r^2 \rangle \propto (p - p_c)^\mu t , \tag{11.22}$$

which should correspond to the functional form from the ansatz:

$$(p - p_c)^\mu t \propto t^{2k} f((p - p_c)t^x) \propto t^{2k}[(p - p_c)t^x]^\mu . \tag{11.23}$$

This results in the exponent relation

$$2k = 1 - \mu x , \tag{11.24}$$

or

$$k = \frac{1 - \mu x}{2} . \tag{11.25}$$

Scaling Behavior in the Limit $p < p_c$ Similarly, we know that the behavior in the limit of $u \ll -1$ should be proportional to $(p_c - p)^{\beta-2\nu}$. Consequently, the scaling ansatz gives

$$(p_c - p)^{\beta-2\nu} \propto t^{2k} f((p - p_c)t^x) \propto t^{2k}[(p_c - p)t^x]^{\beta-2\nu} , \tag{11.26}$$

which results in the exponent relation:

$$2k + x(\beta - 2\nu) = 0 . \tag{11.27}$$

Solving to Find the Exponents We solve the two equations for x and k, finding

$$k = \frac{1}{2}[1 - \frac{\mu}{2\nu + \mu - \beta}] , \tag{11.28}$$

Fig. 11.8 Plots of
$r^2(t; p, L)$ for
$p = 0.45, 0.50, 0.55$ rescaled
according to the scaling
theory

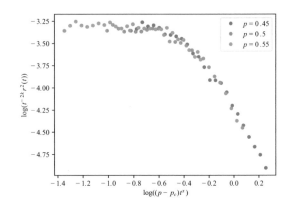

and

$$x = \frac{1}{2\nu + \mu - \beta} \, . \tag{11.29}$$

Our argument therefore shows that the scaling ansatz is indeed consistent with the limiting behaviors we have already determined, and it allows us to make a prediction for k and x.

Testing the Scaling Ansatz We can test the scaling function by a direct plot of the simulated result. The scaling relation states that $r^2(t) = t^{2k} f[(p - p_c)t^x]$, which means that $r^2(t)t^{-2k} = f[(p - p_c)t^x]$. If we therefore plot $r^2(t)t^{-2k}$ on one axis and $(p - p_c)t^x$ on the other axis, all the data for the various values of p should fall onto a common curve corresponding to the function $f(u)$. This is illustrated in Fig. 11.8, which shows that the scaling ansatz is in good correspondence with the data. Indeed, the plot also shows that the assumptions about the shape of the scaling function $f(u)$ are correct.

Interpreting the Dimension of the Walk at $p = p_c$ When $p = p_c$, we find that

$$\langle r^2 \rangle \propto t^{2k} = t^{\frac{2\nu - \beta}{2\nu + \mu - \beta}} \, , \tag{11.30}$$

We can write this relation in the same way as we wrote the behavior of an ordinary random walk,

$$t \propto r^{d_w} \, , \tag{11.31}$$

where d_w is the dimension of the random walk. We have therefore found that

$$d_w = \frac{1}{k} = 2 + \frac{\mu}{\nu - \frac{\beta}{2}} \, , \tag{11.32}$$

which is a number larger than 2. This means that for a given time, the walk remains more compact, which is consistent with our intuition.

Defining the Cross-Over Time We have introduced a cross-over time, t_0, which is defined so that

$$(p - p_c)t_0^x \simeq 1 \,, \tag{11.33}$$

which gives

$$t_0 \propto |p - p_c|^{-1/x} \propto |p - p_c|^{-(2v+\mu-\beta)} \,. \tag{11.34}$$

Interpreting the Crossover Time How can we interpret this relation? We could decompose the relation to be:

$$t_0 \propto \frac{|p - p_c|^{\beta-2v}}{|p - p_c|^\mu} \,, \tag{11.35}$$

where we know that the average radius of gyration for clusters are

$$[R_s^2] \propto |p - p_c|^{\beta-2v} \,, \tag{11.36}$$

This gives us an interpretation of the cross-over time for diffusion:

$$t_0(p) \propto \frac{[R_s^2]}{D} \,, \tag{11.37}$$

where D is the diffusion constant. Why is this time not proportional to ξ^2/D, the time it takes to diffuse a distance proportional to the correlation length? The difference comes from the particular way we devised the experiment: the walker was dropped onto a randomly selected occupied site.

Interpreting the Behavior for $p > p_c$ Let us now address what happens when $p > p_c$. In this case, the contributions to the variance of the position has two main terms: one term from the spanning cluster and one term from the finite clusters.

$$[\langle r^2 \rangle] = Dt = \frac{P}{p}D't + R_s^2 \,, \tag{11.38}$$

where the first term, $P/pD't$ is the contribution from the random walker on the infinite cluster. This term consists of the diffusion constant D' for a walker on the spanning cluster, and the prefactor P/p which comes from the probability for the walker to land on the spanning cluster: For a random walker placed randomly on an occupied site in the system, the probability for the walker to land on the spanning cluster is P/p, and the probability to land on any of the finite clusters is $1 - P/p$.

The second term is due to the finite cluster. This term reaches a constant value for large times t. The only time dependence is therefore in the first term, and we can write:

$$Dt = \frac{P}{p}D't , \qquad (11.39)$$

for long times, t. That is:

$$D' = \frac{Dp}{P} \propto (p - p_c)^{\mu - \beta} \propto \xi^{-\frac{\mu - \beta}{\nu}} \propto \xi^{-\theta} . \qquad (11.40)$$

where we have introduce the exponent

$$\theta = \frac{\mu - \beta}{\nu} . \qquad (11.41)$$

Interpreting the Crossover Time for $p > p_c$ We have therefore found an interpretation of the cross-over time t_0, and, in particular for the appearance of the β in the exponent. We see that the cross-over time is

$$t_0 \propto \frac{|p - p_c|^{\beta - 2\nu}}{|p - p_c|^{\mu}} \propto \frac{\xi^2}{D'} . \qquad (11.42)$$

The interpretation of t_0 is therefore that t_0 is the time the walker needs to travel a distance ξ when it is diffusing with diffusion constant D' on the spanning cluster.

Diffusion on the Spanning Cluster

How does the random walker behave on the spanning cluster? We have found that for $p > p_c$ and for $t > t_0$ the mean square displacement increases according to

$$\langle r^2 \rangle = D't \propto (p - p_c)^{\mu - \beta}t , \qquad (11.43)$$

For $t < t_0$, we expect the behavior to be

$$\langle r^2 \rangle \propto t^{2k'} , \qquad (11.44)$$

as illustrated in Fig. 11.6.

Interpretation of t_0 **for Walks on the Spanning Cluster** We expect the relations to be valid up to the point (t_0, ξ^2), where both descriptions should provide the same result. Therefore we expect

$$\xi \propto t_0^{2k'} \propto D't_0 , \qquad (11.45)$$

and therefore that

$$t_0 \propto \frac{\xi^2}{D'} \propto \frac{(p - p_c)^{-2v}}{(p - p_c)^{\mu - \beta}} \propto (p - p_c)^{-(2v + \mu - \beta)} . \qquad (11.46)$$

Consequently, the value of t_0 is the same for diffusion on the spanning cluster as for diffusion on any cluster including the spanning cluster. In general, we can interpret t_0 as the time it takes for the walker to diffuse to the end of the cluster when $p < p_c$, and the time it takes to diffuse to a distance ξ on the spanning cluster when $p > p_c$.

Interpretation of k for Walks on the Spanning Cluster   Let us check the other exponent, k'. We find that

$$\xi^2 \propto (p - p_c)^{-2(2v + \mu - \beta)k'} , \qquad (11.47)$$

and therefore that

$$k' = \frac{v}{2v + \mu - \beta} , \qquad (11.48)$$

which is not the same as we found in (11.28) for all clusters. We find that k' is slightly larger than k.

Interpretation of k' and k   What is the interpretation of k'? If we consider random walks on the spanning cluster only, the behavior at $p = p_c$ is described by

$$\langle r^2 \rangle \propto t^{2k'} , \qquad (11.49)$$

this gives

$$r^{1/k'} \propto t \propto r^{d_w} , \qquad (11.50)$$

where d_w can be interpreted as the dimension of the random walk. For the case of random walkers on the spanning cluster at $p = p_c$ we have therefore found that4

$$d_w = 2 + \frac{\mu - \beta}{v} . \qquad (11.51)$$

The fractal dimension is larger than 2. This corresponds to the walker getting stuck on the percolation cluster, and the structure of the walk is therefore more dense or compact.

The Diffusion Constant D

We can use the theory we have developed so far to address the behavior of the diffusion constant with time. Fick's law can generally be formulated as

$$\langle r^2 \rangle = \mathscr{D}t , \tag{11.52}$$

or, equivalently, we can find the diffusion constant for Fick's law from:

$$\mathscr{D} = \frac{\partial}{\partial t} \langle r^2 \rangle . \tag{11.53}$$

Now, we have established that for diffusion on the spanning cluster for $p = p_c$, the diffusion is anomalous. That is, the relation between the square distance and time is not linear, but a more complicated power-law relationship

$$\langle r^2 \rangle \propto t^{2k'} . \tag{11.54}$$

As a result, we find that the diffusion constant $\mathscr{D}'$ for diffusion on the spanning cluster defined through Fick's law is

$$\mathscr{D}' \propto \frac{\partial}{\partial t} t^{2k'} \propto t^{2k'-1} . \tag{11.55}$$

We can therefore interpret the process as a diffusion process where $\mathscr{D}$ decays with time.

In the anomalous regime, we find that

$$r \propto t^{k'} , \tag{11.56}$$

and therefore that

$$r^{1/k'} \propto t . \tag{11.57}$$

We can therefore also write the diffusion constant $\mathscr{D}'$ as

$$\mathscr{D}' \propto t^{2k'-1} \propto r^{2-1/k'} \propto r^{-\theta} . \tag{11.58}$$

We could therefore also say that the diffusion constant is decreasing with distance. The reverse is also generally true: Whenever $\mathscr{D}$ depends on the distance, we will end up with anomalous diffusion.

We can rewrite the dimension, d_w, of the walk to make the relation between the random walker and the dimensionality of the space on which it is moving more obvious:

$$d_w = 2 - d + \frac{\mu}{\nu} + d - \frac{\beta}{\nu} , \qquad (11.59)$$

where we recognize the first term as

$$\tilde{\zeta}_R = 2 - d + \frac{\mu}{\nu} , \qquad (11.60)$$

and the second term as the fractal dimension, D, of the spanning cluster:

$$D = d - \frac{\beta}{\nu} . \qquad (11.61)$$

We have therefore established the relation

$$d_w = \tilde{\zeta}_R + D . \qquad (11.62)$$

This relation is actually generalizable, so that for a random walker restricted to only walk on the backbone, the dimension of the walker is

$$d_{w,B} = \tilde{\zeta}_R + D_B . \qquad (11.63)$$

Exercises

Exercise 11.1 (Random Walks on the Spanning Cluster) In this exercise we will use and modify the program `percwalk` from the text to study random walks in percolation systems, and on the spanning cluster in particular. We want to find the dimension d_w of a two-dimensional random walk on the spanning cluster.

(a) Find the distance $\langle r^2 \rangle$ as a function of the number of steps N for random walks on the spanning cluster for $p = p_c$.
(b) Find the dimension, d_w of the walk, from the relation $\langle r^2 \rangle \propto N^{2/d_w}$.
(c) Find the distribution $P(r, N)$ for the position r as a function of the number of steps N for a random walker on the percolation cluster.
(d) (Advanced) Can you produce a data-collapse for the distribution $P(r, N)$.
(e) (Advanced) Can you determine the functional form of the distribution $P(r, N)$. Is it a Gaussian?

Exercise 11.2 (Random Walks on Percolation Clusters) In this exercise we will use and modify the program `percwalk` to study random walks on the spanning cluster of a percolation system.

(a) Find the distance $\langle r^2 \rangle$ as a function of the number of steps N for random walks on the spanning cluster for $p < p_c$ and for $p > p_c$.

(b) Plot $\log \langle R^2 \rangle$ as a function of N for various values of p.

(c) Can you find the behavior of the correlation length ξ from this plot?

(d) Discuss the behavior of the characteristic cross-over time t_0 based on the plot.

Exercise 11.3 (Self-Avoiding Walks on Fractals) (Advanced) In this exercise we will use the program `percwalk` to study a self-avoiding random walker on the spanning cluster. In this exercise you will need to collect extensive statistics to be able to determine the scaling behavior.

(a) Find the distance $\langle R^2 \rangle$ as a function of the number of steps N for random walks on the spanning cluster for $p = p_c$.

(b) Find the dimension, d_w of the walk, from the relation $\langle R^2 \rangle \propto N^{2/d_w}$.

Dynamic Processes in Disordered Media **12**

In this chapter we start to address dynamical processes that generate percolation-like disordered structures. We will address the scaling behavior of diffusion fronts for diffusion in disordered media, and develop a theory for the front width. We will address the slow displacement of fluids in porous media with the invasion-percolation model. For this model, we will map the model onto a percolation system and demonstrate how the model changes with the introduction of an additional aspects, gravity.

12.1 Introduction

So far, we have studied the behavior and properties of systems with *disorder*, such as a model porous material in the form of a percolation system. That is, we have studied properties that depend on the existing disorder of the material. In this chapter, we will start to address dynamical processes that generate percolation-like disordered structures, but where the structures evolve, develop, and change in time.

The first dynamic problem we will address is the formation diffusion fronts, and we will demonstrate that the front of a system of diffusing particles can be described as a percolation system.

The second dynamic problem we will address is the slow displacement of one fluid by another in a porous medium. We will in particular demonstrate that the invasion percolation process generates a fractal structure similar to the percolation cluster by itself - it is a process that drives itself to a critical state, similar to the notion of Self-Organized Criticality [3]. We will then address how we can study similar processes in the gravity field, and, in particular, the influence of stabilizing and destabilizing mechanisms.

© The Author(s) 2024
A. Malthe-Sørenssen, *Percolation Theory Using Python*, Lecture Notes
in Physics 1029, https://doi.org/10.1007/978-3-031-59900-2_12

12.2   Diffusion Fronts

The first dynamical problem we will address is the structure of a diffusion front [31] on a square lattice. One example of such a process is the two-dimensional diffusion of particles from a source at $x = 0$ into the $x > 0$ plane, when particles are not allowed to overlap. The system of diffusing particles is illustrated in Fig. 12.1.

Exact Solution   For this problem we know the exact solution for the concentration, $c(x, t)$, of particles, corresponding to the occupation probability $P(x, t)$. The solution to the diffusion equation with a constant concentration at one boundary, here $P(x = 0, t) = 1$, is the error function, which is the integral over a Gaussian function:

$$P(x, t) = 1 - \text{erf}(\frac{x}{\sqrt{Dt}}) , \tag{12.1}$$

where the error function is defined as the integral:

$$\text{erf}(u) = \frac{2}{\sqrt{2\pi}} \int_0^u e^{\frac{-v^2}{2}} dv . \tag{12.2}$$

This solution produces the expected variance $\langle x^2 \rangle = Dt$, where D is the diffusion constant for the particles. There is no y (or z) dependence for the solution.

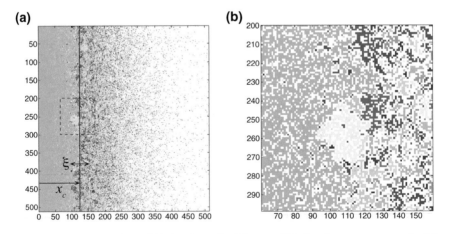

Fig. 12.1 Illustration of the diffusion front. Particles are diffusing from a source at the left side. We address the front separating the particles connected to the source from the particles not connected to the source. (**a**) The average distance is given by x_c shown in the figure. The width of the front, $\xi(x_c)$, as x_c is also illustrated. The different clusters are colored to distinguish them from each other. The close-up in figure (**b**) illustrates the finer details of the diffusion fronts and the local cluster geometries

Structure of Clusters Connected to Diffusion Front We will address the structure of connected clusters of diffusing particles. Two particles are connected if they are neighbors so that they inhibit each others diffusion in a particular direction. If we fix t, we notice that the system will be compact close to $x = 0$, and that there only will be a few thinly spread particles when $x \gg \sqrt{Dt}$. In this system, the occupation probability varies with both time t and spatial position x. However, we expect the system of diffusing particles to be connected to the source out to a distance x_c corresponding to the point where the occupation probability is equal to the percolation threshold p_c for the lattice type studied. That is:

$$P(x_c, t) = p_c , \qquad (12.3)$$

defines the center of the diffusion front: the front separating the particles that are connected to the source from the particles that are not connected to the source. We notice that $x_c(t) = \sqrt{Dt}$.

Width of the Diffusion Front What is the width of the diffusion front? For a given time t, the occupation probability decreases with distance from x_c. The correlation length will therefore depend on the distance $\delta x = x - x_c$ to x_c. We expect that a cluster may be connected to the front if it is within a distance ξ of x_c. Particles that are further away than the local correlation length, ξ, will not be connected over such distances, and will therefore not be connected. Particles that are closer to x_c than ξ will typically by connected through some connecting path. We will therefore introduce ξ as the width of the front, corresponding to the distance at which the local correlation length, due to the occupation probability $P(x, t)$, is equal to the distance from x_c. The local correlation length at a position x, $\xi(x)$, is given as

$$\xi(x) = \xi_0 |P(x, t) - p_c|^{-\nu} , \qquad (12.4)$$

The distance w at which $\xi(x_c + w) = w$ gives the width of the front. We can write this self-consistency equation for w as

$$w = \xi_0 |P(x + w, t) - p_c|^{-\nu} . \qquad (12.5)$$

Let us introduce a Taylor expansion of $P(x)$ around $x = x_c$:

$$P(x, t) \simeq P(x_c, t) + \left. \frac{dP}{dx} \right|_{x_c} (x - x_c) , \qquad (12.6)$$

where we recognize that $x_c \propto \sqrt{Dt}$ gives

$$\left. \frac{dP}{dx} \right|_{x_c} \propto \frac{1}{\sqrt{Dt}} \propto \frac{1}{x_c} . \qquad (12.7)$$

We insert this into the self-consistency Eq. 12.5) getting

$$w = \xi_0 |w \left. \frac{dP}{dx} \right|_{x_c} |^{-\nu} \propto (w/x_c)^{-\nu} , \tag{12.8}$$

which gives

$$w \propto x_c^{\nu/(1+\nu)} . \tag{12.9}$$

The width of the front therefore scales with the average position of the front, and the scaling exponent is related to the scaling exponent of the correlation length for the percolation problem.

Time Development   What happens in this system with time? Since x_c is increasing with time, we see that the relative width decreases:

$$\frac{w}{x_c} \propto \frac{x_c^{\nu/(1+\nu)}}{x_c} \propto x_c^{-\frac{1}{1+\nu}} . \tag{12.10}$$

This effect will also become apparent under renormalization. Applying a renormalization scheme with length b, will result in a change in the front width by a factor $b^{\nu/(1+\nu)}$, but along the y-direction the rescaling will simply be by a factor b. Successive applications will therefore make the front narrower and narrower.

12.3   Invasion Percolation

We will now study the slow injection of a non-wetting fluid into a porous medium saturated with a wetting fluid. In the limit of infinitely slow injection, this process is termed *invasion percolation* for reasons that will soon become obvious [12, 38].

Physical System: Fluid Saturated Porous Medium   When a non-wetting fluid is injected slowly into a saturated porous medium, the pressure in the non-wetting fluid must exceed the capillary pressure in a pore-throat for the fluid to propagate from one pore to the next, as illustrated in Fig. 12.2. The pressure difference, δP needed corresponds to the capillary pressure P_c, given as

$$P_c = \frac{\Gamma}{\epsilon} , \tag{12.11}$$

where Γ is the interfacial surface tension, and ϵ is the characteristic size of the pore-throats in the porous medium. However, there will be some disorder present in the porous medium corresponding to local variation in the characteristic pore sizes ϵ. This will lead to a distribution of the capillary pressure thresholds, P_c, needed to invade a particular pore. We will assume that the medium can be described as

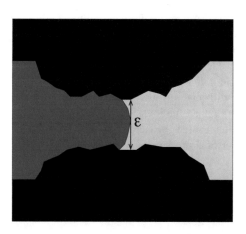 

Fig. 12.2 Illustration of the invasion percolation process in which a non-wetting fluid is slowly displacing a wetting fluid. The left figure shows the interface in a pore throat: the pressure in the invading fluid must exceed the pressure in the displaced fluid by an amount corresponding to the capillary pressure $P_c = \Gamma/\epsilon$, where Γ is the interfacial surface tension, and ϵ is a characteristic length for the pore throat. The right figure illustrates the invasion front after injection has started. The fluid may invade any of the sites along the front indicated by small circles. The site with the smallest capillary pressure threshold will be invaded first, changing the front and exposing new boundary sites

a set of pores connected with pore throats with a uniform distribution of capillary pressure thresholds, and we will assume that the capillary pressure thresholds are not correlated but statistically independent. We can therefore rescale the pressure scale, by subtracting the minimum pressure threshold and dividing by the range of pressure thresholds, and describe the system as a matrix of critical pressures P_i required to invade a particular site.

Modeling the Fluid Displacement Process The fluid displacement process can be modeled by assuming that all the sites on the left side of the matrix are in contact with the invading fluid. The pressure in the invading fluid is increased slowly, until the fluid invades the connected site with the lowest pressure threshold. This generates a new set of invaded sites in contact with the inlet, and a new set of neighboring sites. The invasion process continues until the invading fluid reaches the opposite side. Further injection will then not produce any further fluid displacement, the fluid will flow through the system through the open path generated.

Computational Implementation How can we transfer this model description to a computational model? We introduce a lattice of pores to represent the pore throat sizes. For each lattice site, there is a critical pore size into that pore, with a critical pressure, p_i, needed to push the fluid into this pore. We map the pressure onto a scale from 0.0 to 1.0, where 1.0 represents the pressure needed to invade all pores in the lattice.

We then start to gradually increase the pressure in the fluid and allow the fluid to invade from the left side of the lattice. Let us say we have increased the pressure to the value p ($0 \le p \le 1$). This would mean that all sites that have $p_i \le p$ *and* that are connected to the left side would be invaded.

This corresponds to a percolation problem. If we make a percolation system with occupation probability p, then the fluid will have invaded all the clusters that are connected to the left side. Thus, we have mapped the invasion percolation problem onto a percolation problem. Let us implement this approach.

First, we generate a random lattice of critical pressures and an array of pressures p that we will loop through:

```
import numpy as np
import matplotlib.pyplot as plt
from scipy.ndimage import measurements
L = 400
z = np.random.rand(L,L) # Random distribution of thresholds
p = np.arange(0.0,0.7,0.01)
```

We step gradually through this set of p-values, finding the clusters of connected sites that have p-values smaller than p[npstep]

```
for nstep in range(len(p)-1):
    zz = z<p[nstep]
    lw,num = measurements.label(zz)
```

Then, we find the labels of all the clusters on the left side of the lattice. All the clusters with these labels are connected to the left side and are part of the invasion percolation cluster called cluster. We do this in two steps. First, we find a list of unique labels that are on the left side. Then we find all the clusters with labels that are in this list using the numpy-function isin:

```
    leftside = lw[:,0]
    leftnonzero = leftside[np.where(leftside>0)]
    uniqueleftside = np.unique(leftnonzero)
    cluster = np.isin(lw,uniqueleftside)
```

Then we make a matrix that stores at what time t (pressure value $p(t)$) a particular site was invaded. This is done by simply adding a 1 for all set sites at time t to a matrix pcluster. The first clusters invaded will then have the highest value in the pcluster matrix. We use the pcluster matrix to visualize the dynamics.

```
    pcluster = pcluster + 1.0*cluster
```

Finally, we check if the fluid has reached the right-hand side by comparing the labels on the left-hand side with those on the right-hand side. If any labels are the same, there is a cluster connecting the two sides (a spanning cluster), and the fluid invasion process stops:

```
# Check if it has reached the right hand side
span = np.intersect1d(lw[:,1],lw[:,-1])
if (len(span[np.where(span > 0)])>0):
    break
```

The whole program for the simulation, including initialization of `pcluster` is then:

```
# Example program for studying invasion percolation problems
# NOTE: This is not an optimal but an educational algorithm
import numpy as np
import matplotlib.pyplot as plt
from scipy.ndimage import measurements
L = 400
p = np.arange(0.0,0.7,0.01)
z = np.random.rand(L,L) # Random distribution of thresholds
pcluster = np.zeros((L,L),float)
for nstep in range(len(p)-1):
    zz = z<p[nstep]
    lw,num = measurements.label(zz)
    leftside = lw[:,0]
    leftnonzero = leftside[np.where(leftside>0)]
    uniqueleftside = np.unique(leftnonzero)
    cluster = np.isin(lw,uniqueleftside)
    pcluster = pcluster + 1.0*cluster
    # Check if it has reached the right hand side
    span = np.intersect1d(lw[:,1],lw[:,-1]) # Test perc
    if (len(span[np.where(span > 0)])>0): break
plt.imshow(np.log(pcluster),origin="lower")
```

Results for Fluid Displacement Process The resulting pattern of injected nodes is illustrated in Fig. 12.3, where the colors indicate the pressure at which the injection took place. It can be seen from the figure that the injection occurs in bursts. When a site is injected, many new connected neighbors are available as possible sites to invade. As the pressure approaches the pressure needed to percolate to the other side, these newly appearing sites of the front will typically also be invaded, and invasion will occur in gradually larger regions. These bursts have been characterized by Furuberg et al. [12], and it can be argued that the distribution of burst sizes as well as the time between bursts are power-law distributed.

Mapping Invasion Percolation onto Percolation Based on this algorithmic model for the fluid displacement process, it is also easy to relate the invasion percolation problem to ordinary percolation. For an injection pressure of p, all sites with critical pressure below or equal to p are in principle available for the injection process. However, it is only the clusters of such sites connected to the left side that will

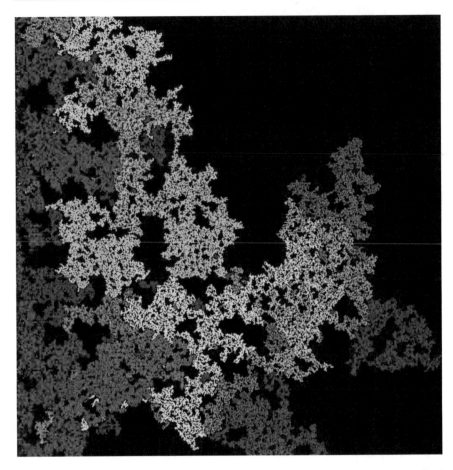

Fig. 12.3 Illustration of the invasion percolation cluster. The color-scale indicates normalized pressure at which the site was invaded

actually be invaded, since the invasion process requires a connected path from the inlet to the site for a site to be filled. We will therefore expect that the width of the invasion percolation front corresponds to the correlation length $\xi = \xi_0(p_c - p)^{-\nu}$ as p approaches the percolation threshold p_c, because this is the length at which clusters are connected. That is, clusters that are a distance ξ from the left side will typically be connected to the left side, and therefore connected, whereas clusters that are further away than ξ will typically not be connected and therefore not invaded. This shows that the critical pressure will correspond to p_c. This also shows that when the fluid reaches the opposite side, the system is exactly at p_c, and we expect the invasion percolation cluster to have the same scaling properties as the spanning cluster at $p = p_c$. There will be small differences, because the invasion percolation cluster also contains smaller clusters connected to the left side, but we do not expect these to change the scaling behavior of the cluster. That is, we expect the fractal dimension of the invasion percolation cluster to be D. This implies that the density of the displaced fluid decreases with system size.

Invasion Percolation with and Without Trapping The process outlined above does, however, not contain all the essential physics of the fluid displacement process. For displacement of an incompressible fluid, a region that is fully bounded by the invading fluid cannot be invaded, since the displaced fluid does not have any place to go. Instead, we should study the process called invasion percolation with trapping. It has been found that when trapping is included, the fractal dimension of the invasion percolation cluster is slightly smaller [10]. In two dimensions, the dimension is $D \simeq 1.82$.

This difference between the process with and without trapping disappears for three-dimensional geometries because trapping become unlikely in dimensions higher than 2. Indeed, direct numerical modeling shows that the fractal dimension for both the ordinary percolation system and invasion percolation is $D \simeq 2.5$ for invasion percolation with and without trapping.

Gravity Stabilization

The invasion percolation cluster displays self-similar scaling similar to that of ordinary percolation. This implies that the position $h(x, p)$ of the fluid front as a function of the non-dimensional applied pressure p is given as the correlation length—since this is how far clusters connected to the left side typically are connected. That is, when p approaches p_c, the average position of the front is $\bar{h}(x, p) = \xi(p) = \xi_0(p_c - p)^{-\nu}$. The width, $w(p)$ of the front is also given as the correlation length:

$$w(p) = \xi_0(p_c - p)^{-\nu} , \qquad (12.12)$$

as p approaches p_c both the front position and the front width diverges, that is, both the front position $\bar{h}$ and the width, w, are proportional to the system size L:

$$\bar{h} \propto w \propto L , \qquad (12.13)$$

However, when the system size increases, we would expect other stabilizing effects to become important. For a very small, but finite fluid injection velocity, the viscous pressure drop will eventually become important and comparable to the capillary pressure. Also, for any deviation from a completely flat system for a system with a slight different in densities, the effect of the hydrostatic pressure term will eventually become important. We will now demonstrate how we may address the effect of such a stabilizing (or destabilizing) effect [6, 25].

Invasion Percolation in a Gravity Field Let us assume that the invasion percolation occurs in the gravity field. This implies that the pressure needed to invade a pore depends both on the capillary pressure, and on a hydrostatic term. The pressure P_i^c needed to invade site i at vertical position x_i in the gravity field is:

$$P_i^c = \frac{\Gamma}{\epsilon} + \Delta\rho g x_i \,, \tag{12.14}$$

We can again normalize the pressures, resulting in

$$p_i^C = p_i^0 + \frac{\Delta\rho g}{\Gamma\epsilon^2} x_i' \,, \tag{12.15}$$

where the coordinates are measured in units of the pore size, ϵ, which is the unit of length in our system. The last term is called the Bond number:

$$Bo = \frac{\Delta\rho g}{\Gamma\epsilon^2} \,, \tag{12.16}$$

Here, we will include the effect of the bond number in a single number g, so that the critical pressure at site i is:

$$p_i^c = p_i^0 + g x_i' \,, \tag{12.17}$$

where p_i^0 is a random number between 0 and 1.

Computational Implementation   We implement this by changing the values of the pressure threshold p_i in the computational code:

```
g = 0.001
grad = g*np.meshgrid(range(L),range(L))[0]
z = z + grad
```

The whole program then becomes

```
# Now we add the effect on gravity - modifying the values of z
import numpy as np
import matplotlib.pyplot as plt
from scipy.ndimage import measurements
L = 400
p = np.arange(0.0,0.7,0.01)
z = np.random.rand(L,L) # Random distribution of thresholds
g = 0.001
grad = g*np.meshgrid(range(L),range(L))[0]
z = z + grad
pcluster = np.zeros((L,L),float)
for nstep in range(len(p)-1):
    zz = z<p[nstep]
    lw,num = measurements.label(zz)
    leftside = lw[:,0]
```

```
leftnonzero = leftside[np.where(leftside>0)]
uniqueleftside = np.unique(leftnonzero)
cluster = np.isin(lw,uniqueleftside)
pcluster = pcluster + 1.0*cluster
# Check if it has reached the right hand side
span = np.intersect1d(lw[:,1],lw[:,-1]) # Test perc
if (len(span[np.where(span > 0)])>0):
    break
plt.imshow(np.log(pcluster),origin="lower")
```

Visualization of Results The resulting invasion percolation fronts for various numbers of g is illustrated in Fig. 12.4. How can we understand the gradual flattening of the front as g increases from g?

Front Width Analysis This problem is similar to the diffusion front problem. For an applied pressure p the front will typically be connected up to an average distance x_c given as

$$p = p^0 + x_c g \, . \tag{12.18}$$

The front will also extend beyond the average front position. The occupation probability at a distance a from the front is $p' = p_c - ag$, since fewer sites will be set beyond the front due to the stabilizing term g. A site at a distance a is connected to the front if this distance a is shorter to or equal to the correlation length for the occupation probability p' at this distance. The maximum distance a for which a site is connected back to the front therefore occurs when

$$a = \xi(p') = \xi_0(p_c - p')^{-\nu} \, . \tag{12.19}$$

This gives

$$a = \xi(p') = \xi_0(p_c - p')^{-\nu} = \xi_0(p_c - (p_c - ag))^{-\nu} = \xi_0(ag)^{-\nu}a \, , \tag{12.20}$$

and

$$a \propto g^{-\nu/(1+\nu)} \, , \tag{12.21}$$

We leave it as an exercise to find the form of the position $h(p, g)$, and the width, $w(p, g)$, as a function of p and g. We observe that the width has a reasonable dependence on g. When g approaches 0, the width diverges. This is exactly what we expect since the limit $g = 0$ corresponds to the limit of ordinary invasion percolation.

This discussion demonstrates a general principle that we can use to study several stabilizing effects, such as the effect of viscosity or other material or process parameters that affect the pressure needed to advance the front. The introduction of a finite width or characteristic length ξ that can systematically be varied in order

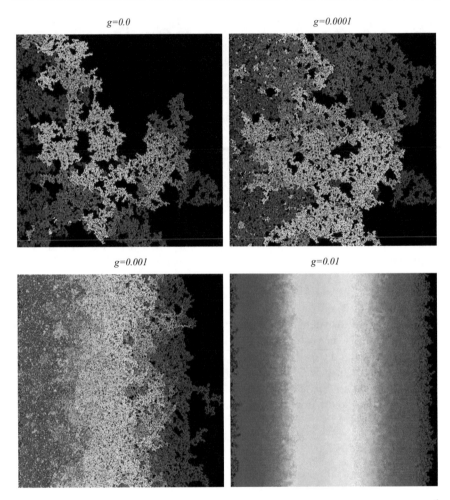

Fig. 12.4 Illustration of the gravity stabilized invasion percolation cluster for $g = 0$, $g = 10^{-4}$, $g = 10^{-3}$, and $g = 10^{-2}$. The color-scale indicates the normalized pressure at which the site was invaded

to address the behavior of the system when the characteristic length diverges is also a powerful method of both experimental and theoretical use.

Gravity Destabilization

The gravity destabilized invasion percolation process corresponds to the case when a less dense fluid is injected at the bottom of a denser fluid. This is similar to the process known as secondary migration, where the produced oil is migrating up through the sediments filled with denser water. Examples of the destabilizing front is shown in Fig. 12.5.

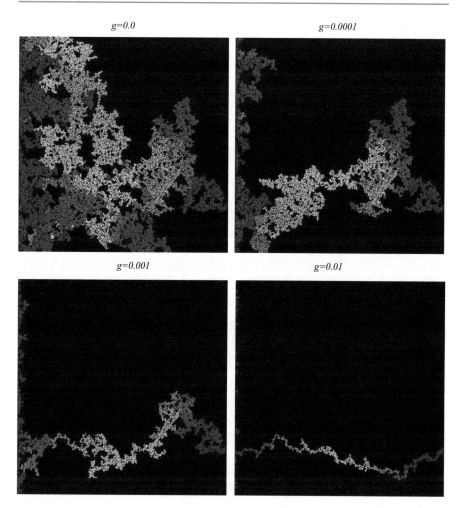

Fig. 12.5 Illustration of the gravity de-stabilized invasion percolation cluster for $g = 0$, $g = -10^{-4}$, $g = -10^{-3}$, and $g = -10^{-2}$. The color-scale indicates normalized pressure at which the site was invaded

We can make a similar argument for the case when $g < 0$, but in this case the front is destabilized, and the correlation length $\xi \propto (-g)^{-\nu/(1+\nu)}$ corresponds to the width of the finger extending front the front. The extending finger can be modeled as a sequence of blobs of size ξ extending from the flat surface. This implies that the region responsible for the transport of oil in secondary migration is essentially a one-dimensional structure: lines with a finite width w. The amount of hydrocarbons left in the sediments during this process is therefore negligible. However, there will be other effects, such as the finite viscosity and the rate of production compared to the rate of flow, which will induce more than one finger. Gravity destabilized invasion percolation is used as a modeling tool in studies of petroleum plays and commercial software packages are available for such simulation.

References

1. J. Adler, Y. Meir, A. Aharony, A.B. Harris, L. Klein, Low-concentration series in general dimension. J. Stat. Phys. **58**(3), 511–538 (1990)
2. A. Aharony, Y. Gefen, A. Kapitulnik, Scaling at the percolation threshold above six dimensions. J. Phys A Math. General **17**(4), L197–L202 (1984)
3. P. Bak, *How Nature Works: The Science of Self-Organized Criticality* (Copernicus, Göttingen, 1996)
4. D.J. Bergman, Y. Kantor, Critical properties of an elastic fractal. Phys. Rev. Lett. **53**(6), 511–514 (1984)
5. H.A. Bethe, Statistical theory of superlattices. Proc. R. Soc. Lond. A **150**, 552–575 (1935)
6. A. Birovljev, L. Furuberg, J. Feder, T. Jssang, K.J. Mly, A. Aharony, Gravity invasion percolation in two dimensions: Experiment and simulation. Phys. Rev. Lett. **67**(5), 584–587 (1991)
7. J.L. Cardy, Introduction to theory of finite-size scaling, in *Current Physics–Sources and Comments*. Finite-Size Scaling, vol. 2 (Elsevier, Amsterdam, 1988), pp. 1–7
8. K. Christensen, N.R. Moloney, *Complexity and Criticality* (Imperial College Press, London, 2005)
9. P.G. de Gennes, La percolation: Un concept unificateur. La Recherche **7**, 919 (1976)
10. M.M. Dias, D. Wilkinson, Percolation with trapping. J. Phys. A Math. General **19**(15), 3131–3146 (1986)
11. S. Feng, P.N. Sen, Percolation on elastic networks: new exponent and threshold. Phys. Rev. Lett. **52**(3), 216–219 (1984)
12. L. Furuberg, J. Feder, A. Aharony, T. Jøssang, Dynamics of invasion percolation. Phys. Rev. Lett. **61**(18), 2117–2120 (1988)
13. Y. Gefen, A. Aharony, S. Alexander, Anomalous diffusion on percolating clusters. Phys. Rev. Lett. **50**(1), 77–80 (1983)
14. G.R. Grimmett, *Percolation*. Grundlehren der mathematischen Wissenschaften, 2nd edn. (Springer, Berlin/Heidelberg, 1999)
15. S. Havlin, D. Ben-Avraham, Diffusion in disordered media. Adv. Phys. **36**(6), 695–798 (1987)
16. H.J. Herrmann, H.E. Stanley, The fractal dimension of the minimum path in two- and three-dimensional percolation. J. Phys. A Math. General **21**(17), L829–L833 (1988)
17. D.C. Hong, H.E. Stanley, Cumulant renormalisation group and its application to the incipient infinite cluster in percolation. J. Numer. Math. **16**(14), L525–L529 (1983)
18. A. Hunt, R. Ewing, B. Ghanbarian, *Percolation Theory for Flow in Porous Media*. Lecture Notes in Physics, 3rd edn. (Springer, Berlin, 2014)
19. L.P. Kadanoff, Scaling laws for ising models near ${T}_{c}$. Phys. Phys. Fizika **2**(6), 263–272 (1966)
20. Y. Kantor, I. Webman, Elastic properties of random percolating systems. Phys. Rev. Lett. **52**(21), 1891–1894 (1984)

© The Author(s) 2024
A. Malthe-Sørensen, *Percolation Theory Using Python*, Lecture Notes in Physics 1029, https://doi.org/10.1007/978-3-031-59900-2

21. H. Kesten, *Percolation Theory for Mathematicians*. Progress in Probability (Birkhäuser, Basel, 1982)
22. P.R. King, M. Masihi, *Percolation Theory in Reservoir Engineering* (World Scientific (Europe), London, 2018)
23. S. Kirkpatrick, Percolation and conduction. Rev. Mod. Phys. **45**(4), 574–588 (1973)
24. B.J. Last, D.J. Thouless, Percolation theory and electrical conductivity. Phys. Rev. Lett. **27**(25), 1719–1721 (1971)
25. P. Meakin, A. Birovljev, V. Frette, J. Feder, T. Jøssang, Gradient stabilized and destabilized invasion percolation. Phys. A Stat. Mech. Appl. **191**(1), 227–239 (1992)
26. Y. Meir, R. Blumenfeld, A. Aharony, A.B. Harris, Series analysis of randomly diluted nonlinear resistor networks. Phys. Rev. B **34**(5), 3424–3428 (1986)
27. S. Roux, Relation between elastic and scalar transport exponent in percolation. J. Numer. Math. **19**(6), L351–L356 (1986)
28. M. Sahimi, Relation between the critical exponent of elastic percolation networks and the conductivity and geometrical exponents. J. Phys. C Solid State Phys. **19**(4), L79–L83 (1986)
29. M. Sahimi, J.D. Goddard, Elastic percolation models for cohesive mechanical failure in heterogeneous systems. Phys. Rev. B Condens. Matter Mater. Phys. **33**(11), 7848–7851 (1986)
30. M. Sahini, M. Sahimi, *Applications Of Percolation Theory* (Wiley Press, Hoboken, 2003)
31. B. Sapoval, M. Rosso, J.F. Gouyet, The fractal nature of a diffusion front and the relation to percolation. J. Phys. Lett. **46**(4), 149–156 (1985)
32. W. Sierpinski, Sur une courbe dont tout point est un point de ramification. Comp. Rend. Acad. Sci. Paris **160**, 302–305 (1915)
33. S. Solomon, G. Weisbuch, L. de Arcangelis, N. Jan, D. Stauffer, Social percolation models. Phys. A Stat. Mech. Appl. **277**(1), 239–247 (2000)
34. H.E. Stanley, Cluster shapes at the percolation threshold: and effective cluster dimensionality and its connection with critical-point exponents. J. Phys. A Math. General **10**(11), L211–L220 (1977)
35. H.E. Stanley, *Introduction to Phase Transitions and Critical Phenomena*, Reprint edn. (Oxford University Press USA, New York, 1987).
36. H.E. Stanley, A. Coniglio, Fractal structure of the incipient infinite cluster in percolation, in *Percolation Structures and Processes*, ed. by G. Deutsher, R. Zallen, J. Adler (Adam Hilger, Bristol, 1983)
37. D. Stauffer, A. Aharony, *Introduction To Percolation Theory*, 2nd edn. (Wiley Press, Hoboken, 1992)
38. D. Wilkinson, J.F. Willemsen, Invasion percolation: a new form of percolation theory. J. Numer. Math. **16**(14), 3365–3376 (1983)
39. K.G. Wilson, Renormalization group and critical phenomena. I. Renormalization group and the kadanoff scaling picture. Phys. Rev. B Condens. Matter Mater. Phys. **4**(9), 3174–3183 (1971)
40. J.G. Zabolitzky, D.J. Bergman, D. Stauffer, Precision calculation of elasticity for percolation. J. Stat. Phys. **44**(1), 211–223 (1986)

Index

© The Author(s) 2024

A. Malthe-Sørenssen, *Percolation Theory Using Python*, Lecture Notes in Physics 1029, https://doi.org/10.1007/978-3-031-59900-2

Printed in the United States
by Baker & Taylor Publisher Services